Water Resources Development and Management
Series Editors: Asit K. Biswas and Cecilia Tortajada

Editorial Board

Dogan Altinbilek (Ankara, Turkey)
Chennat Gopalakrishnan (Honolulu, USA)
Jan Lundqvist (Linköping, Sweden)
Alexandra Pres (Feldafing, Germany)
Anthony Turton (Pretoria, South Africa)
Olli Varis (Helsinki, Finland)

Asit K. Biswas · Cecilia Tortajada · Rafael Izquierdo (Eds.)

Water Management in 2020 and Beyond

With 29 Figures and 30 Tables

Prof. Asit K. Biswas
Third World Centre for Water Management
Avenida Manantial Oriente 27
Los Clubes, Atizapan
Estado de México, 52958
Mexico
akbiswas@thirdworldcentre.org

Rafael Izquierdo
Water Institute of Aragon
Capitán Portolés, 1-3-5, 8 floor
50004 Zaragoza
Spain
rizquierdoa@aragon.es

Cecilia Tortajada
International Centre for Water and
Environment (CIAMA)
Water Institute of Aragon
Capitán Portolés, 1-3-5, 8 floor
50004 Zaragoza
Spain
ceda@aragon.es

ISBN 978-3-540-89345-5   e-ISBN 978-3-540-89346-2

DOI 10.1007/978-3-540-89346-2

Water Resources Development and Management ISSN: 1614-810X

Library of Congress Control Number: 2008939846

© Springer-Verlag Berlin Heidelberg 2009

This work is subject to copyright. All rights are reserved, whether the whole or part of the material is concerned, specifically the rights of translation, reprinting, reuse of illustrations, recitation, broadcasting, reproduction on microfilm or in any other way, and storage in data banks. Duplication of this publication or parts thereof is permitted only under the provisions of the German Copyright Law of September 9, 1965, in its current version, and permission for use must always be obtained from Springer. Violations are liable to prosecution under the German Copyright Law.

The use of general descriptive names, registered names, trademarks, etc. in this publication does not imply, even in the absence of a specific statement, that such names are exempt from the relevant protective laws and regulations and therefore free for general use.

Cover design: deblik, Berlin

Printed on acid-free paper

9 8 7 6 5 4 3 2 1

springer.com

This book is dedicated to the memory of two eminent development experts, Morris Miller of Canada and Hans Pfeifer of Germany, who participated in the preparation of this book. The world is much poorer because of their departure and we have lost two close friends whose memory we treasure.

Foreword

Aragon: Forum for Global Debates on Water

Water is such a scarce resource, and of such high value, that it merits profound reflection on its future management. With this in mind, the idea emerged of organising a unique international summit in Aragon – because of the issue involved and because it would be the first to which top-level experts from a variety of sectors and disciplines would be invited to debate the future of water management.

Accordingly, in November 2006, the newly established International Centre of Water and the Environment (CIAMA-La Alfranca) of the Government of Aragon brought together leading national and international experts on water issues, as well as on population, development, economics, agriculture, energy, health, and technology, for a workshop entitled "Water Management Beyond 2020." Various probable scenarios for the management of water in the medium and long term were debated. Those who also took part in organising the event included the Mexico-based Third World Centre for Water Management, headed by its President Prof. Asit K. Biswas, who is also a Senior Advisor to CIAMA, and the Sasakawa Peace Foundation of the United States, with its President Mr. Keiji Iwatake. I wish to thank them, as well as Dr. Cecilia Tortajada, Scientific Director of the CIAMA and President of the International Water Resources Association, for their dedication and effort in conceiving the event and ensuring its success.

In forums such as these, Aragon shows why it takes a leading role in the international debate on water, climate change, and sustainable development, as highlighted by the recently completed international water exhibition, Expo Zaragoza 2008. The people of Aragon have always wanted to go further in broadening the perspective with which we view water management, in the knowledge that water is the primordial element in guaranteeing sustainable socioeconomic development of our society. Our presidency of the Environmental Conference of the European Regions (ENCORE) is a landmark in that leadership vocation. We can also show concrete practical examples that reflect our collective commitment. For example, we have launched the largest regional plan in all of Europe for the treatment of wastewater of the region to ensure high quality of water in our rivers, lakes, and aquifers, with an investment of more than 1 billion euros and a pioneering formula of public–private participation. Furthermore, we are acquiring vital experience in models of social participation, thanks to the Aragon Water Commission, a forum without equal in the world.

The present book offers a wonderful opportunity to reflect in-depth on the past, the present, and, above all, the future. A window has been opened to the year 2020. We can gaze through it in the hope that, in the years that remain till we reach that horizon, we shall find the best formulae for the responsible management of water.

Zaragoza, Spain

Alfredo Boné-Pueyo,
Minister of Environment,
Government of Aragon, Spain

Preface

As the eminent Danish physicist and Nobel Laureate Niles Bohr had noted, "Prediction is very difficult, especially of the future". While forecasting the future reliably is an extremely complex and difficult process, one issue can be predicted with complete certainty: the world beyond 2020 will be vastly different to what it is today. However, what is not possible to prophesise with any degree of certainty is how the world will differ beyond 2020 compared to what it is today.

The changes we are likely to see during the next two decades are likely to be far ranging and far reaching, and these changes will come concurrently from different parts of the world, from different sectors and disciplines, from academic and business communities, from NGOs and civil society as a whole, and from changing social attitudes and perceptions.

There will be many drivers of change, some known but others unknown. Even for the known drivers, it will be very difficult to predict their overall impacts in terms of the timings of their onsets, magnitudes, and spatial distributions. Making reliable forecasts will be further complicated due to many other factors: speed, extent, and impacts of globalisation and free trade; governance of institutions (national, regional, and international, as well as public, private, and civil society organisations) and the efficiency of their functioning; formulation and implementation of national and international policies; acceleration of information and communication revolution; advances in knowledge; changes in public perceptions and expectations, etc.

The water sector is an integral component of the global system, and thus, it will not be immune to these changes. On the contrary, it will be influenced in a variety of ways by the future global changes. Many of these changes will originate from non-water sectors, or from non-water-related issues, on which the water profession will have no, or at best limited, say or control. This will make water management beyond 2020 an exceedingly complex task with a complexity that is highly likely to increase with time.

In spite of the rhetoric of the past two decades that business-as-usual is no longer an option for efficient water management in the future, the fact remains that the profession in general believes, at least implicitly, that future changes will be mostly incremental. Consequently, current practices need to be gradually improved to take care of the expected future changes. In other words, business-unusual solutions may not be necessary, perhaps because most may believe that such responses cannot be developed within a medium-term timeframe.

Customarily, water professionals have mostly ignored the global forces that are somewhat external to the water sector, even though these are likely to shape water use, availability patterns (both in terms of quantity and quality), and management practices in the future in many significant ways. For example, water professionals continue to ignore the various water-related implications of globalisation and free trade, even though these developments are likely to unleash forces that may have radical impacts on water use, availability, and requirement patterns in many different countries of the world, irrespective of their current levels of economic development or sophistication in water management and use of latest technological advances. Some of these impacts are already visible in numerous countries, ranging from the United States to Japan and from China to Mexico. The impacts of these unleashed forces are only likely to increase in the future.

Current and potential impacts of globalisation and free trade on the water sector have been consistently ignored by the water profession, as well as national and international institutions that deal with water. Nor has there been any attempt to discuss them seriously at major international water events like World Water Fora or Stockholm Water Symposia. Furthermore, there are no visible signs that this situation, although unacceptable, is likely to change in any significant manner in the foreseeable future. Instead, considerable emphasis continues to be placed on "same old stuff" (SOS) like integrated water resources management and integrated river basin management: paradigms that have yet to be operationalised after more than half a century of rhetoric and expenditure of hundreds of million of dollars. While considerable emphasis is still being placed on the issues that have long passed their "sell by" dates, the profession continues to neglect, or even totally ignore, major developments in areas like biotechnology, desalination, information and communication, population trends, aging, and immigration, even though developments in these areas are likely to affect the water security of many nations in the *near* future.

There is no question that the future water-related issues of the world are likely to be very different from those that were witnessed in the past or are being encountered at present. While historical knowledge and past experience are always useful, identifying, analysing, and solving water problems of the future will require new insights, additional skills, innovative approaches, adaptable mindsets, and proactive institutions. It will also require a holistic approach that can successfully coordinate energy, food, environment, and industrial policies of a nation, all of which have intimate linkages to water. Each will affect the others and, in turn, will be affected by the others. These will not be zero-sum games: considerable compromises and trade-offs have to be made, based not only on techno-economic and environmental terms but also on sociopolitical considerations, which may often vary from one country to another or even from one region to another in the same country.

Since the world is changing very rapidly, water management practices and processes beyond 2020 must change as well. Status quo, with only incremental changes, that is generally preferred by most institutions and professions, is no longer an option. The nature and type of future water problems should be objectively and carefully analysed in the light of expected global changes from inside the water sector as well as outside the sector, which are likely to affect it. Past forecasts and recent trends may no longer shed any meaningful light on the coming, new, turbulent environment in the world of water, which will somehow have to accommodate diversified, sometimes even opposing, requirements reflecting different needs and interests of various stakeholders, political processes, and institutional requirements. The situation will be further complicated by rapid technological changes, accelerated globalisation, and relentless economic competition between countries and within countries, all of which are likely to have direct and indirect water-related ramifications.

In this future-oriented book, some of the world's leading authorities discuss many of the opportunities and challenges that the water profession may be expected to face beyond the year 2020. Many of the driving forces that are likely to affect water management beyond 2020 will still be traditional, like population and urbanisation. However, their implications are likely to be very different to what have been witnessed in the past, or are being observed at present. There will be new challenges due to non-traditional drivers like globalisation, free trade, HIV/AIDS, climate change, technological developments in areas like desalination and biotechnology, and increasing intersectoral linkages between water, food, energy, and environmental securities, which will make efficient and equitable water management more complex than ever before. These and other related issues are discussed at some length in this book.

As we move further into the twenty-first century, a clear vision of the future and how this vision can be achieved to serve humanity better will be needed. It will be important to discern what societies value most, which in turn will determine and drive their unique visions of the future. Taking these steps in a timely and cost-effective and socially acceptable manner will not be an easy task, especially as the precedents available at present are very few and somewhat restricted in their scopes. However, it will be an essential requirement for the water profession in the future.

In our view, there is now a revolution taking place in the area of water management, even though most observers may not be aware of it. In the wake of this revolution, long-held concepts and models of water management are likely to undergo accelerated evolution, and some are likely to disappear altogether. Never before in the history of water have such profound changes taken place within such a short timeframe as we are likely to witness during the next two or three decades. It will thus be necessary to determine how best our water systems can be planned and managed so that they benefit society as a whole very significantly and on a long-term basis. This will be a major challenge, perhaps even the most difficult and complex challenge the water profession has had to face. Nevertheless, it is a challenge that the water profession must meet successfully.

In order to initiate a global dialogue on water management beyond 2020, the International Centre of Water and Environment (CIAMA-La Alfranca), in cooperation with the Water Institute of Aragon, Third World Centre for Water Management of Mexico, International Water Resources Association, and Sasakawa Peace Foundation of the United States, invited 30 leading international experts from various water-related fields to meet in Zaragoza to discuss future-oriented issues associated with water management. The present book is based on the invited papers that were very specifically commissioned for this invitation-only workshop. We are especially grateful to Mr. Fernando Otal, Secretary General of the Water Institute of Aragon, for all his support that made the Zaragoza workshop possible.

According to an African proverb, tomorrow belongs to the people who prepare for it today. We earnestly hope that the process that was initiated in Zaragoza was the beginning of a process that will help the water profession to understand, define, and appreciate the nature and type of water problems of the future. At the end of the process, we hope we shall have a series of cost-effective, socially acceptable, and environmentally friendly solutions to well-defined water problems of the future.

Atizapan, Mexico *Asit K. Biswas*
Zaragoza, Spain *Cecilia Tortajada*
Zaragoza, Spain *Rafael Izquierdo*

Contents

Changing Global Water Management Landscape 1
Asit K. Biswas and Cecilia Tortajada

More Urban and More Aged: Demographic Pressures to Global Water Resources by 2050 35
Olli Varis

Water and the Next Generation – Towards a More Consistent Approach .. 65
Malin Falkenmark

Adaptive Water Management: Strengthening Laws and Institutions to Cope with Uncertainty 89
Carl Bruch

In Search of a Comprehensive Approach to Sustainable Management of Water Resources in the World Community 115
Kazuo Takahashi

Science, Ideology and Sustainable Development: An Actor-Oriented Approach 125
Peter Söderbaum

Leading and Managing Change in Water Reform Processes Capacity Building Through Human Resource Development 137
Hans Pfeifer

The European Water Framework Directive: Potential for Change and Implications Beyond 2020 149
José Albiac and Juan Ramón Murua

Towards a Climate-Proof Netherlands 165
Michiel van Drunen, Aalt Leusink and Ralph Lasage

The Strategic Role of Water in Alleviating the Human Tragedy Associated with HIV/AIDS in Africa 195
Jeanette Rascher, Peter Ashton and Anthony Turton

Irrigation and Water Policies in Aragon 213
José Francisco Aranda-Martín

Singapore Water: Yesterday, Today and Tomorrow 237
Teng Chye Khoo

The Pace of Change in Seawater Desalination by Reverse Osmosis 251
Ian Lomax

Index 259

Contributors

Albiac, José Senior Researcher, Department of Agricultural Economics, Agrifood Research and Technology Centre (CITA-DGA), Av. Montañana 930, 50059 Zaragoza, Spain.

Aranda-Martin, José Francisco Head, Department of Planning and Development, Water Institute of Aragon, Capitán Portolés, 1-3-5, 9 floor, 50005 Zaragoza, Spain.

Ashton, Peter Principal Researcher and Divisional Fellow, Council for Scientific and Industrial Research (CSIR) – Natural Resources and the Environment, PO Box 395, Pretoria, 0001, South Africa.

Biswas, Asit President, Third World Centre for Water Management, Av. Manantial Oriente 27, Los Clubes, Atizapán, Estado de México, 52958, Mexico.

Bruch, Carl Senior Attorney, Co-Director, International Programs, Environmental Law Institute, 2000 L Street NW, Suite 620, Washington, DC 20036, USA.

Falkenmark, Malin Professor, Stockholm International Water Institute (SIWI), Drottninggatan 33, SE-111 51, Stockholm, Sweden.

Izquierdo, Rafael Director, Water Institute of Aragon, Capitán Portolés, 1-3-5, 8 floor, 50004 Zaragoza, Spain.

Khoo, Teng Chye Chief Executive, Public Utilities Board of Singapore, 111 Somerset Road No. 15-01 Singapore 238164, Republic of Singapore.

Leusink, Aalt Director, LOASYS, Management & Advice in Water and Environment, Groen van Prinstererlaan 34, 2271 En Voorburg, the Netherlands.

Lomax, Ian Manager Large Projects, FILMTEC (tm) Membranes, Dow Water Solutions, Dow Deutschland Anlagengesellschaft mbH. PO Box 20, D-77834 Rheinmunster, Germany.

Murua, José Ramón Associate Professor, Department of Applied Economics V, University of the Basque Country (UPV/EHU), Ave. Lehendakari Agirre 83, 48015 Bilbao, Spain.

Pfeifer, Hans (Late) InWEnt, Director, Department of Environment, Wielinger Strasse 52, 82340 Feldafing, Germany.

Ralph Lasage Researcher, Institute for Environmental Studies, Vrije Universiteit, De Boelelaan 1087, 1081 HV Amsterdam, the Netherlands.

Rascher, Jeanette Research Group Leader, Council for Scientific and Industrial Research (CSIR) – Natural Resources and the Environment, Water Resource Governance Systems, PO Box 395, Pretoria, 0001, South Africa.

Soderbaum, Peter Professor Emeritus, Department of Economics, Malardalens University, Box 883, 72123 Vasteras, Sweden.

Takahashi, Kazuo Professor, Division of International Studies International Christian University, 3-10-2 Osawa Mitaka, Tokyo 181-8585, Japan.

Tortajada, Cecilia Scientific Director, International Centre for Water and Environment (CIAMA), Water Institute of Aragon, Capitán Portolés, 1-3-5, 8 floor, 50004 Zaragoza, Spain.

Turton, Anthony Unit Fellow, Council for Scientific and Industrial Research (CSIR) – Natural Resources and the Environment, Water Resource Competence Area, PO Box 395, Pretoria, 0001, South Africa.

Van Drunen, Michiel Researcher, Institute for Environmental Studies, Vrije Universiteit, De Boelelaan 1087, 1081 HV Amsterdam, the Netherlands.

Varis, Olli Professor, Helsinki University of Technology, Water Resources Laboratory, PO Box 5200, 02015 Espoo, Finland.

Changing Global Water Management Landscape

Asit K. Biswas and Cecilia Tortajada

> *Change is the law of life. And those who look only to the past or present are certain to miss the future.*
>
> John F. Kennedy

Introduction

In the sixteenth century, the eminent Renaissance scholar Leonardo Da Vinci said that water is the driver of nature. During his lifetime, some may have considered this to be an overstatement, but some half a millennium later, Leonardo's understanding of the role, relevance and importance of water to society and nature can be considered to have been prophetic. Water is increasingly considered to be the lifeblood of the planet and, at present, it will certainly not be an overstatement to claim that without efficient water management, the future social and economic development of the world would be seriously constrained, or even significantly jeopardised. Both developed and developing countries will require implementation of more and more efficient water management policies and practices in terms of both quantity and quality. However, developing countries will need to improve their water management practices and processes much more than developed countries, especially as the current practices of the former have significant potential for improvement.

As the human population has grown, the global food requirements have increased as well and with it the requirement to water needed to produce the necessary food. At present, most of the global water use is accounted for by the agricultural sector, estimated at about 70%. Historically, water–food interrelationships have always been important. However, in recent years, these linkages have become more and more complex because of national and international politics, social and environmental considerations, different forms and levels of subsidies and tariffs applied to food products in different countries, globalisation, free trade, changing standards of living, institutional and legal requirements, technological developments and management practices.

A.K. Biswas (✉)
Third World Centre for Water Management, Av. Manantial Oriente 27, Los Clubes, Atizapán, Estado de México, 52958, Mexico
e-mail: akbiswas@thirdworldcentre.org

In percentage terms, agricultural water use worldwide has been declining in recent decades (in absolute terms it is still increasing). However, industrial water use has been steadily increasing since the advent of the Industrial Revolution, when global development patterns changed dramatically. Industrial water requirements started to increase rapidly, as did the need for collection, treatment and environmentally safe disposal of wastewater generated. As the nations industrialised, their industrial water requirements increased as well. Accordingly, in all the industrialised countries, industrial water demands have long exceeded domestic water needs. Similarly, as developing countries become industrialised, their water requirements for this sector have increased as well. Accordingly, industrial water needs in many developing countries at present are already higher than their domestic water requirements, and this trend is likely to accelerate in the future.

With rapid industrialisation, improving standards of living and consistent demands for a better quality of life, energy requirements of nations have gone up as well. For example, in recent years, electricity requirements of many developing countries like Brazil, China, India, the Philippines, South Africa, Thailand and Turkey have been increasing at more than 5% per year. Since these demands are increasing from a historically low base, they will continue to rise at a high rate during the coming decades. This comparatively recent development has had major water-related implications, since no large-scale generation of electricity can be achieved without water. Water, of course, is essential for hydropower generation. Equally, thermal and nuclear power cannot be generated without the availability of large quantities of cooling water. In turn, the water sector is a major user of energy, especially in terms of pumping. These symbiotic interrelationships between the water and energy sectors, for the most part, have been ignored by the water and energy professionals in nearly all the countries of the world.

Environmental and social implications of development projects started to become important during the 1970s. They gathered further momentum during the 1980s and 1990s, by which time environmental impact assessments of all major projects had become mandatory in most countries. During the past three decades, many lessons have been learnt as to how major development projects can be planned, constructed and managed so that the overall results are economically efficient, socially acceptable and environmentally friendly. Water development projects have been no exception to this comparatively new but essential requirement, also working towards the generation of new knowledge for better planning and management practices.

All the above-mentioned and many other associated developments have ensured that, over the past five to six decades, water planning and management have become an increasingly complex and difficult task all over the world. Many of the current and emerging trends indicate that this process will become even more complex in the coming years, most likely at a more accelerated rate than has been witnessed during the recent historical past. Because of these accelerated complexities that should be realistically expected in the future, many of which are likely to stem from non-water sectors on which water professionals are likely to have a limited say or control, it is essential that water planners and managers should start thinking seriously about the trends and developments in the post-2020 world. Changes in many water-related areas are likely to be so fast in the future that past experiences and present practices may at best be of limited use to manage them in an efficient and timely manner.

Planning and construction of large water development projects often span more than two decades. This means that many new generations of water projects that are being planned now may become fully operational around 2030, or even later, when the overall environment and the boundary conditions within which they will be operating are likely to be very different from what exist at present. Thus, it is essential that the water professionals start taking a futuristic look so that the major planning assumptions of current projects are still likely to be valid in a post-2030 world, when they are likely to become operational. Regrettably, this is an aspect that is being mostly ignored at present.

While it will be a very difficult task to predict the future changes that will affect water management in a post-2020 world with any degree of certainty in terms of their spatial and temporal nature, their magnitudes or even the timing of the onset of any specific change, one aspect can be predicted with complete certainty: that is that major changes will occur which will require development of new paradigms for water management, including new and innovative approaches for water planning. Yesterday's experience and today's approaches are likely to be of limited value in understanding or identifying tomorrow's trends, let alone solving the day after tomorrow's water problems. For water managers, business unusual has to be the new *mantra* of the day.

Future Drivers of Change

During the next two to three decades, there will be many drivers of change which will affect water availability and use patterns. These driving forces can be classified into three general categories. First, some of these drivers are well known since they have affected water management practices and processes in many ways in the past. However, even for these drivers, like population and urbanisation, their potential implications in the future are likely to be somewhat different than have been witnessed in the past. Linear extrapolation of past trends will give a misleading development scenario for the future. Hence, even for this group of drivers, considerable thinking and further research will be needed to predict reliably their future water-related implications and how these can be efficiently incorporated into water management processes.

There are other drivers, like economic growth and energy generation, which constitute the second category that will affect patterns of water use and consumption, as well as wastewater generation. While the relevance of these factors has sometimes been implicitly recognised in water management, they have seldom been explicitly considered for policy and planning purposes.

The third category of drivers is those which, for the most part, are being completely ignored by mainstream water professionals at present. Among these drivers are issues like globalisation, free trade, immigration, advances in biotechnology and desalination, diseases like HIV/AIDS, changing management paradigms and evolving social attitudes and perceptions.

Only some of the above-mentioned drivers will be discussed in this introductory chapter, primarily to give an indication as to how these issues are likely to affect water management in the future, directly or indirectly. Many of these drivers have been discussed in more detail by other authors in this book. It should be noted that the changing patterns of these drivers and their potential implications to the water sector are still not fully understood and thus, not surprisingly, even less appreciated by water professionals at present. In fact, a good research agenda in this overall area is currently conspicuous by its absence. Equally, the impacts and the relevance of these drivers may vary from one country to another and may even vary within any specific country, as well as over time.

Economic Growth and Water

It has been anecdotally known for at least half a century that economic growth affects water consumption patterns through a variety of pathways, which can be direct, indirect or tertiary. Some of these changes are predictable and quantifiable and thus comparatively easy to handle within the existing management process. However, many other changes are unknown, unpredictable and/or intangible, as a result of which it has been very difficult to incorporate them explicitly in terms of overall planning. Equally, water resources, if planned and managed properly, can act as an engine for promoting regional economic growth and for alleviating poverty. Comprehensive analyses carried out by the Third World Centre for Water Management clearly indicate that projects like Bhakra Nangal in India (Rangachari, 2006), High Aswan Dam in Egypt (Biswas and Tortajada, 2009) and Ataturk Dam in Turkey (Tortajada, 2004) have radically transformed the economic growth patterns of the regions concerned by generating new employments, enhancing energy and food security conditions and improving major social indicators.

The last 20 years have seen radical changes in the economic growth patterns of several developing countries. The recession of 2008 and its expected continuation in 2009 will undoubtedly affect the earlier forecasts of economic growths in all parts of the world, but the relative patterns of economic developments are unlikely to change significantly from what were estimated earlier, especially over the medium term. All these changing economic growth patterns will, in turn, have a myriad of implications in terms of water planning and management through a series of interacting pathways and feedback loops. These interacting patterns are still not fully understood, but they would invariably have noteworthy implications for both water quantity and quality management in the future.

The countries having some of the highest prospects for economic growth in the coming years are Brazil, Russia, India and China, collectively known as BRICs, a term that was first coined by Goldman Sachs. The aggregate contribution of BRICs to global growth has been around 30% since 2000 and exceeded that of the United States by the end of 2008. China has already become the world's fourth largest economy and is expected to surpass Germany by around 2010. The GDP of China has more than doubled since 2000, meaning that it has effectively created a contribution that is equivalent to an additional France, two Canadas or three Indias. In the case of India, the increase in terms of its GDP is equivalent to that of the Netherlands in 2000 (Goldman Sachs, 2006a).

Globally, the pace of economic growth is expected to slow between 2008 and 2010. However, long-term high-growth projections pose potential major challenges to the environment by placing additional pressure on natural resources utilisation. In the case of water, these economic growth rates will have major implications in terms of access, quantity, quality, equity and management and investment requirements. These developments will pose major challenges in the future because current policy measures will not be able to address these emerging developments adequately or systematically. Unless a determined effort is made to change the current approaches properly, current and future water policy measures are unlikely to address the expected changes adequately. There are no signs at present that innovative and appropriate sets of water policy measures can be formulated and implemented in a timely manner in countries that are likely to have high economic growth rates.

The type of economic growth that can help preserve the environment, and decrease pollution, water included, should also improve the living conditions of the average citizen significantly. The growth should improve income distributions within the countries as well as between the countries. Nevertheless, even in a scenario of robust economic growth, increased income and improved environmental quality are not always related since more affluent countries and better-off citizens do not automatically protect the environment. At least for the OECD countries, in general, pollution levels have decreased, and the interest in preserving the environment has increased in response to domestic demands that can be supported by sound investment and public policy initiatives (OECD, 2007).

Since developing countries started from a lower base, Goldman Sachs (2006b) has estimated that three conglomerates of countries (G6[1], BRICs and N11[2]) would together need, over the next 5 years, total investments of approximately $4 trillion for infrastructural developments. Almost half of this requirement is likely to be accounted for by the BRICs. Even if the above-estimated amounts are actually invested, it would still take the BRICs another 25 years to have a level of infrastructure equivalent to that of the G6 countries at present.

In order for the above-mentioned countries to achieve real growth, Goldman Sachs (2006b) identified certain priority issues which have to be considered, such as savings, population growths and productivity gains. They also identified issues such as good governance, sensible policies, strong institutions and appropriate legal and regulatory frameworks to encourage the participation of the private sector. All countries should consider these as priority issues irrespective of their stages of development or the development sectors concerned.

Goldman Sachs has estimated that commitment from the private sector to energy, telecommunications, transport and water in all developing countries over the past

---

[1] G6 consists of United States, Japan, Germany, Italy, United Kingdom and France.

[2] N11 consist of Bangladesh, Egypt, Indonesia, Iran, Korea, Mexico, Nigeria, Pakistan, Philippines, Turkey and Vietnam.

15 years has been approximately $60 billion per year. In the short term, telecommunications (mostly mobiles) may receive 50% of the total global private investments, while electricity generation, roads and water projects may receive 35, 15 and 5%, respectively. Over time, investments are expected to be channelled to electricity generation and roads, although this will depend on the policy frameworks of the different countries and their attractiveness to the private sector (Goldman Sachs, 2006b).

An interesting aspect that should be noted is that, during the second half of the 1990s, there was optimism that water supply and wastewater management problems of the urban developing world would be solved by the private sector, which would bring with it additional investment funds and good management expertise that the public sector did not have. After some early promise, the multinational companies have basically lost much of their initial enthusiasm to invest in the water sector during the post-2000 period. Thus, an important aspect of any future private sector involvement in the water sector has to be the separation of the short-term blips from the long-term trends.

Population

Historically, population has been an important driver for the water sector. As the global population increases, unless the existing water management practices become more and more efficient, the world will require significant additional quantities of water to sustain its inhabitants at a reasonable standard of living. However, unlike oil or coal, water is a renewable resource. This means that water can be used, and then reused, many times as long as proper water quality management practices are followed. Thus, water poses a very different type of management approach compared to nonrenewable resources like oil or coal.

In terms of water management, population has many other implications, in addition to just numbers. Among these are concentration (urban or rural), age structure and gender-related issues.

Historically, the world population has increased steadily. However, the growth rates have regularly varied with time. In 1950, the global population was estimated at 2.53 billion (Table 1). By 2000, it had more than doubled to 6.12 billion (an increase of 1.42 times). By 2050, half a century later, it is estimated to increase to 9.19 billion. This indicates an estimated increase of around 50%, which is only about one-third of the growth that was witnessed during the previous 50-year period.

The decline in the global population growth rates has been quite significant in the recent past. For example, during 1970–1975, the growth rate was 2.02% per year. This is expected to decline by 82% during the 2045–2050 period, to about 0.36%.

While the world population will increase steadily up to 2050, its distribution over the different regions will not be homogenous. While Asia's share during the 1950–2050 period will remain fairly similar in percentage terms, Africa's share will increase from 8.8 to 21.7%. In contrast, Europe's share will decline from 21.6% to 7.2% (Table 2).

Table 1 Total population (thousands), 1950–2050

Year	Total population
1950	2,535,093
1960	3,031,931
1970	3,698,676
1980	4,451,470
1990	5,294,879
2000	6,124,123
2010	6,906,558
2020	7,667,090
2030	8,317,707
2040	8,823,546
2050	9,191,287

Source: Adapted from United Nations (2007a)

The main increases in global population during the 2005–2050 period are expected to occur in only eight countries. In order of population increment, these are likely to be India, Nigeria, Pakistan, Congo, Ethiopia, the United States, Bangladesh and China. Together they may account for almost half of the expected increase in global population during this period.

In the coming years, life expectancy is expected to increase and fertility rates are likely to decrease. Migration, legal or illegal, will become an increasingly important consideration, certainly significantly more important than whatever was witnessed in the recent past. For example, during 2005–2010, net migration is expected to be higher than natural increase (births minus deaths) in countries as diverse as Belgium, Canada, Luxemburg, Singapore, Spain, Sweden and Switzerland. During the same period, China, India, Indonesia, Mexico and the Philippines are expected to witness the highest rates of net emigration (United Nations, 2007a).

Another interesting trend is the increase in middle-income class (annual income levels between $6,000 and $30,000) of 2 billion people by 2030. This is expected increase pressure on and competition for local and global resources. By 2050, the middle-income class is expected to include six of the N11 countries (Egypt, Philippines, Indonesia, Iran, Mexico and Vietnam) and three of the BRICs (China, India and Brazil). These countries are likely to contribute to approximately 60% of

Table 2 Population distribution by regions, 1950, 2007 and 2050 (in percentage)

	1950	2007	2050
Africa	8.8	14.5	21.7
Asia	55.6	60.4	57.3
Europe	21.6	11.0	7.2
Northern America	6.6	8.6	8.4
Latin America and the Caribbean	6.8	5.1	4.8
Oceania	0.5	0.5	0.5

Source: Adapted from United Nations (2007a)

the global GDP (Wilson and Dragusanu, 2008). The impacts of such growth rates could result in a demand for policies for a cleaner environment from an increasingly affluent population. Equally, if countries do not grow as expected and inequalities increase within countries and between regions, the environment may not become a priority concern for the population. If so, neither governments nor private sector institutions may be able to invest in cleaner practices and processes as often anticipated at present.

Ageing Population

An important factor that is now receiving virtually no attention from water managers is the rapid increase in ageing population. Globally, the population of older persons is growing at a rate of 2.6% per year, which is much higher than the annual population growth rate of 1.1%. The older population is expected to continue growing more rapidly than the population in other age groups, at least until 2050. Such rapid growth will require far-reaching economic and social adjustments in most countries (United Nations, 2007b), including significant changes in practices for managing water and other natural resources.

By 2000, the number of elderly people in the world had reached 600 million, which was nearly three times that of 1950. By 2050, this number is expected to reach 2 billion, some 22% of the entire population (corresponding figure in 1950 was only 8%). In developed countries, the number of elderly persons in 2000 was over one-fifth of their entire population. This is expected to increase to nearly one-third by 2050. In developing countries, the corresponding increase between 2000 and 2050 is estimated to be from 8% to 20%. In other words, by 2050, developing countries will have very similar percentages of elderly people as developed countries had in 2000. This will mean that the global median age will increase from 28 years in 2000 to 38 years by 2050. In 2000, Uganda had the youngest median age (15 years) and Japan had the highest (43 years).

The number of people over 80 years is projected to increase by five times, to 402 million, between 2005 and 2050. During this period, the 80-plus population in China will increase from 15.4 million to 103 million; in India from 7.8 million to 51 million; and in the United States from 10.6 million to 31 million. In other words, the 80-plus population is expected to increase at 3.9% per year, which will be more than 3.5 times the overall estimated population growth rate for this period. Thus, all the current forecasts indicate that the elderly will become an important social and political force during the post-2020 world.

Ageing population will affect the patterns of national developments and resource consumptions in a variety of ways. It will affect rates of economic growth, savings, consumptions, government revenues through taxations and expenditures, housing, health-care expenses, pension commitments and intergenerational financial transfers. In addition, compared to the earlier experience from the developed world, developing countries will be ageing at a much faster rate. This will mean that developing countries will have less time to adjust, compared to what was experienced by the developed world earlier, even though the former will have less financial capabilities, less management and administrative capacities and more efficient governance practices compared to what the developed countries had access to earlier when their population aged (Lawson et al., 2005).

The issue of an increasingly elderly population has yet to receive adequate attention in the water profession. Yet it is likely to be an important policy consideration in nearly all developing countries during the next three to four decades. Countries like India and China have at present a major demographic window of opportunity to restructure their economic development activities during the next two to three decades, with a trained, experienced and energetic workforce. However, after 2010, the number of elderly people will start to increase quite rapidly, so much so that by 2030, China will have more elderly people than the current total population of the United States (Varis, 2009).

The steady ageing of populations in East (excluding China) and South (excluding India) Asia and the two most populous Asian countries (China and India) is shown by Varis (2009) in this book. The problem of increasing elderly populations will be a complex one for all developing countries to address properly in the future. It will have major social, economic and political implications and will affect the water sector through many direct and indirect pathways.

The relationship between water management and an increasingly elderly population is completely unexplored at present. It is likely that they will affect each other in a variety of ways, only a few of which will be discussed in this chapter.

First, in the context of rural, semi-urban and peri-urban areas of many developing countries, and in the absence of water and wastewater connections at the household level, people are forced to use communal land and water bodies for hygiene purposes. For a steadily increasing number of elderly people, routine daily hygiene practices will become a serious chore, especially when physical movements become difficult or when they are sick. With improvements in health care, education and nutrition, people will be living for increasingly longer periods. Absence of water and wastewater collection facilities at home will pose particular burdens on an increasing elderly population, as well as on their families.

Second, as the older generations of people retire from work, considerable knowledge, experience and collective memory would be increasingly lost. In a country like Japan, a significant percentage of many knowledgeable and experienced people will retire from the water sector during the next 5–10 years. The overall institutional knowledge and experience levels in the water sector may decline very suddenly, and this cannot be immediately replaced by younger and newer recruits. This loss of institutional expertise and memory has already been identified as a serious issue for the water sector by the concerned ministry in Japan. This situation will become more widespread in the coming decades in both developed and developing countries.

Third, it is generally the young people who migrate to urban areas in search of better standards of living. Thus, the percentages of young people in the rural areas are likely to continue to decline, with attendant decline in the economic, social and cultural activities in the rural areas. This could accelerate the breakdown of the extended family systems. Consequently, the family support that was available to the earlier generations of elderly people would most probably continue to decline steadily. This will contribute to increasing social and economic problems in terms of deteriorating lifestyles of the elderly. It will also increase the social stress on their younger family members who may have migrated to the urban areas for a better economic future.

Fourth, in the developed countries, as the number of retired people increases, they are likely to demand better services from their water utilities. Since they will have time on their hands, they will require more information, transparency and communication from their utilities. Thus, water utilities will have to improve significantly their information and communication services with their consumers in the future. In developing countries, water utilities may have to transform themselves from primarily engineering institutions to being more akin to a service industry. This will be a major institutional transformation, which will require significant changes in the mindsets of managers and policy makers. This transformation will not be an easy task, since it is likely to be resisted by many of the existing staff members who may lose some of their power and privileges. This may create some internal conflicts within the water utilities, at least over the medium term.

Finally, virtually no research has been done on the water and wastewater requirements of the elderly and their interrelationships with water through various social, economic and cultural pathways. Unfortunately, not a single institution anywhere in the world is currently conducting serious and sustained research on these types of emerging water-related issues. These need to be studied diligently in the future.

Urbanisation

In terms of urbanisation, urban population is expected to increase from 3.3 billion in 2007 to 6.4 billion in 2050, almost doubling in a little over four decades. Urban areas will absorb all global population growth expected over the next four decades, as well as part of the rural population due to rural–urban migration and to rural settlements that will become urban centres with the passage of time (United Nations, 2007a). Globally, the rural and urban populations are now roughly in balance. However, the world population is expected to be 70% urban by 2050.

Asia has been behind Latin America in terms of the extent of urbanisation. Accordingly, countries in the Asian region are likely to witness a massive urbanisation process during the next two to three decades. While it is estimated that its rural population will remain almost stationary between now and 2025, the urban population is likely to increase by 60%.

This massive urbanisation, which is unprecedented in the entire Asian history, will present new types of water-related challenges that all developing Asian countries would have to face. The types and magnitudes of these challenges are unlikely to be similar to those experienced at present or those faced in the past. They are likely to be of a completely different character, and some of them may even be counter-intuitive. For example, considerable attention has been paid in recent years to the water and wastewater problems of the megacities (defined to have populations of more than 10 million). While megacities consume the lion's share of national resources and political interest, they represented only 3.7% of the global population in 2000 and are expected to represent about 4.7% by 2015. The percentage of population living in the next level of large cities, between 5 and 10 million, is even less: 2.8% in 2000 and expected to rise to 3.7% by 2015 (Biswas, 2006). This means that, in spite of the rate of urbanisation having increased throughout the last few decades, the highest percentages of the world population do not live in megacities and are not likely to live in megacities in the foreseeable near future. Rather, they will live in medium- or small-size cities where provision of services, infrastructure and investment will be a major challenge in the coming years (Tortajada, 2008a).

Urban centres of 500,000 or less accounted for 24.8% of the global population in 2000 (nearly seven times that of the megacities), a figure which is projected to increase to 27% by 2015. The annual average population growth rate for these smaller urban centres is expected to increase from 23.2% during 1975–2000 (comparable growth rate for the megacities was 5%, that is, less than one-quarter) to 28.2% during 2000–2015, compared to 7.5% for the megacities.

These smaller urban centres have received scant attention from national and international institutions as well as from water and development professionals, even though solving the future water and wastewater problems of these smaller urban centres will require at least as much attention as the megacities, if not more. This is because their water problems are likely to be significantly more difficult to resolve than those of the megacities since they do not have adequate financial and political power and technical and management capabilities to handle their much higher urbanisation rates. In addition, the number of these smaller urban centres is significantly higher than the number of megacities, and thus the management efforts needed for the former will be more complex and challenging compared to the latter. In fact, even though the number of people involved in smaller urban centres is 6.7 times that of the megacities and their growth rates are expected to be 4 times higher, it is a strange anomaly that they are receiving conspicuously less attention from national and international water policy makers. Clearly, unless the present policies and interest change radically, these comparatively smaller urban centres are likely to become major water and wastewater "black-holes" of the future.

Another issue worth noting is the dissimilarities in the rates of urbanisation between the megacities of the developed and developing world. For example, cities like London and New York grew progressively over nearly a century, and their gradual growth enabled them to develop effectively their water and wastewater infrastructure and their management services. In contrast, the growth rates of the megacities of the developing world like Dhaka, Jakarta or Karachi in recent decades have simply been explosive (Fig. 1). They have been unable to cope with their explosive growth rates in terms of providing satisfactory drinking water and wastewater management services. They have invariably found it very difficult to run faster just to stay in the same place.

Many of these megacities of the developing world have managed to provide water to their residents, especially to the reasonably well-off residential areas. However,

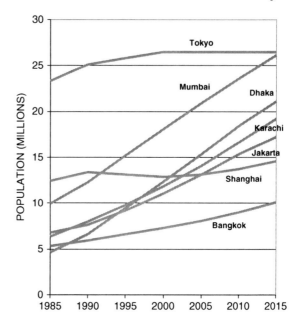

Fig. 1 Population increase in selected Asian megacities

in most cases, the services provided leave much to be desired. For example, supply is often intermittent, and water provided is undrinkable without additional treatment by the households. Furthermore, megacities have progressively fallen behind in the collection, treatment and environmentally safe disposal of wastewater generated. Wastewater may be collected from certain areas of these cities, but they are mostly discharged to nearby rivers, lakes, land or oceans without any treatment or with only primary treatment. Because of this continuing neglect of water quality management, water bodies in and around most urban centres of nearly all developing countries are now heavily contaminated, which has already resulted in serious environmental and health problems for their inhabitants. It is likely that if there is a water crisis in the future, it will not come about because of actual physical scarcity of water, as many have predicted recently, but because of continuing neglect of proper wastewater management practices. Continuation of the present trend will make available water sources increasingly more contaminated and will make provision of clean drinking water more and more expensive, as well as more complex and difficult to manage. The health and environmental costs of continuing mismanagement of wastewater collection, treatment and disposal are likely to increase significantly over the short and medium terms in much of the developing world.

Another noteworthy difference in terms of water management between developed and developing countries is that as the urban centres of the former expanded, their economies were growing as well. Accordingly, it was possible for them to have access to adequate financial resources for efficient urban water and wastewater management. For example, Japan could invest heavily in the construction of urban water infrastructure after 1950 because it was simultaneously experiencing rapid economic growth. Such extensive infrastructure development and major improvements in management practices ensured that unaccounted for water in a megacity like Tokyo could be reduced from an immediate post-war estimate of 90% to about 6% at present, making it one of the most efficient in the world. Equally, cities like Tokyo could invest heavily to control urban flooding, which would have been difficult had Japan's economy not expanded during this period.

In contrast, the rates, extent and complexities of the urbanisation process in the developing world have generally far exceeded the financial and management capacities of the national and the local governments to plan and manage the demographic transition soundly, in terms of providing clean water and wastewater management services to all of the population, efficiently, equitably, sustainably and in a timely manner. The impacts of this inadequately managed urbanisation process are manifested in extensive air, water, land and noise pollution, which is having, and will continue to have, major impacts on human health and quality of life of urban dwellers, as well as imposing major social and economic costs on their respective economies.

Another urbanisation-related problem is the sudden and fast vertical growth, especially in the central business areas, often after decades, or even centuries, of primarily horizontal expansion. This has invariably contributed to a sudden surge in population densities in such areas, with concomitant high water and energy requirements, as well as generation of high waste (wastewater and solid wastes) loads per unit area. The urban centres have simply not been able to cope successfully with such near-instantaneous accelerating demands for water and wastewater management services in the central urban areas. The problem is compounded by the prevailing unsatisfactory water supply and wastewater management services, absence of long-term planning, inadequate management, technical and administrative capabilities, lack of investment funds and high levels of corruption and nepotism.

There are, however, signs of hope. For example, in China, the importance of providing clean drinking water and proper wastewater management services has started to receive increasing high-level political attention. Because its economy has grown very substantially in recent years, the country can financially afford to provide clean water and good wastewater management services to its urban citizens. Water tariffs are rising to meet costs, and this practice has already resulted in reduced industrial water consumptions. Water and wastewater issues have become a priority consideration for national, regional and local policy makers. If such levels of political support and interest continue, it is likely that countries like China will make significant progress in urban water and wastewater management during the coming decades.

Water and Food

Water is an essential requirement for food production. As the world population continues to increase in the coming decades, higher quantities of food will be needed for human and animal consumption. Equally, as developing countries continue to make further economic progress, more people will become affluent. Accordingly, they are likely to change their dietary patterns and demand more protein products, such as meat and milk. This will further increase water requirements because animal husbandry requires more water than crop production.

Such developments do not mean that water demands for producing additional food for a larger and increasingly more affluent global population will increase concomitantly. This is because there is no one-to-one relationship between water requirements and food availability. Crop yields can be increased in a variety of different ways, including more efficient use of fertilisers and pesticides, better-quality seeds and improved management practices. In addition, the food produced should not be the only consideration. In reality, it is the food that is available to the consumers that matters the most. Regrettably, in many developing countries, 25–50% of crops, fruits or vegetables produced at present are not consumed because of heavy losses at every stage of production, transportation, distribution and storage. Reduction of these losses alone will increase food availability very significantly, without any additional water requirements. Accordingly, there are many factors that will affect the total food availability to consumers, and water is not necessarily the only factor, or the most important factor, even though many water professionals have automatically made this erroneous assumption. The numerous factors that are relevant in terms of food production and availability form a complex web of interrelationships, and the problems and their solutions are often location specific. Thus, it is often dangerous, and mostly misleading, to draw generalised conclusions on the quantity of additional water that may be needed to increase the availability of food to consumers in developing countries, without additional comprehensive studies of the overall food systems as a whole, based on reliable and representative data and correct analyses.

Agriculture is by far the major user of water in the world. In many developing countries, agricultural water use accounts for nearly 90% of total water use. However, this percentage has been declining steadily in recent years in most of the developing world, as well as the world as a whole. In contrast, industrial water requirements have been steadily increasing. Nevertheless, in absolute quantitative terms, agricultural water use in most developing countries has been increasing in recent years, and this trend is likely to continue over the medium term.

A major problem with agricultural water use has been that most countries have been pursuing incorrect or inappropriate policies for social, economic and political reasons, many of which are because of fundamental misconceptions. A good example is the energy used for groundwater pumping. Farmers in many countries at present do not pay for the actual volume of groundwater pumped for irrigation. In addition, electricity costs for pumping are very heavily subsidised by many governments. Accordingly, farmers mostly pump more groundwater than is needed for optimising crop production. In turn, this over-pumping is resulting in a steady decline of groundwater levels in many important aquifers all over the world. As the groundwater levels decline, more energy is needed to pump the same quantity of water, which requires additional subsidising of electricity costs. This has contributed to a vicious cycle of overuse of groundwater, declining aquifer levels, increasing losses to the electricity boards and increasing adverse environmental impacts (like land subsidence), none of which are sustainable on a long-term basis. Thus, major policy changes in the water and energy sectors will be needed in the future to balance water and energy uses, stabilise the levels of declining groundwater tables and reduce electricity subsidies to the farmers.

In the future, these types of intersectoral policies need to be carefully analysed, formulated and implemented. Equally, the policies in any specific sector have to be coordinated with the policies in associated sectors. The current and past practices of formulating policies in one sector without adequate consideration of and coordination with the policies in the other sectors will become increasingly costly, inefficient and unsustainable. Herein will lie a major future challenge for the world: how to coordinate appropriately all the concerned resource policies in the areas of water, energy, food and environment; the legal and regulatory frameworks necessary to support these policies; and the coordinating activities of the institutions responsible for formulating and implementing these policies. Such coordination has been very difficult to accomplish in the past because of interinstitutional problems and lack of appreciation of the increasing importance of the need for such coordination. The situation is likely to become even more complex and difficult in the future. Yet this will be an important and critical future requirement that must receive accelerated attention from the governments, research institutions and academia so that appropriate policy instruments can be formulated and implemented for specific situations.

Water and Energy

As the energy needs of the world as a whole continue to increase significantly, the water requirements of the energy sector are likely to increase as well, a fact that has mostly escaped the attention of water and energy planners. Large-scale generation of electricity invariably requires water. Without water, hydropower, an important source of electricity in many countries, cannot be generated. Equally, thermal power generation from coal, oil or natural gas requires significant quantities of cooling water. Nuclear power requires even more cooling water. If the current rate of 5–8% in annual increase in electricity consumption is to be maintained in many developing countries for the indefinite future, as is expected at present, water requirements for the energy sector need to be carefully assessed and then explicitly factored into national water policies. Already, in countries like France, the major user of water is the electricity-generating industry, and not the agricultural or industrial sectors. In the United States, thermoelectric generation represented 39% of all freshwater withdrawals in 2000, which is only a little less than agricultural water requirements (NETL, 2008).

At every stage of the water production and distribution cycle, significant amounts of energy are needed to extract, pump, transport, treat and distribute water to all users. In fact, it is estimated that 2–3% of the world energy consumption is used to pump and treat water for urban residents and industry. In more practical terms, each kilowatt-hour of electricity in the United States requires about 95 L water (E&WR, 2005).

On a global basis, agriculture consumes nearly 70% of available water supplies. There are often tensions between agricultural, domestic and industrial water users, especially in locations where supplies are already insufficient to meet the various demands under existing management practices. Consideration of water withdrawals for the thermonuclear sector, which has been ignored in most countries until now, is likely to increase the tensions further (NIC, 2008).

Use of water in energy and agricultural sectors, leading sometimes to insufficient availability of supply, often suffers from very similar problems: insufficient financial resources, inefficient usage or production, inadequate institutional arrangements, lack of coordination between the sectors, lack of long-term political commitments, inadequate human resources, insufficient community involvement, inappropriate operation and maintenance practices and also provision of insufficient information and communication with their users and consumers (Lawson et al., 2008).

According to the International Energy Agency (IEA, 2007), some $22 trillion of cumulative global investment will be necessary, in 2006 constant dollars, for energy infrastructure over the 2006–2030 period. The investment will be needed to expand supply capacity and to modernise and rehabilitate existing supply facilities. In terms of electricity, more than half of the investments in the industry will be needed for transmission and distribution networks, and the rest for power stations. Since thermoelectric energy generation (including coal, oil, natural gas and nuclear) requires large quantities of reliable cooling water, the estimated investments will require assured supplies of water (NETL, 2008).

Global economic growth will increase demand for energy, both from traditional and non-traditional sources. Globally, economic output, as measured by GDP, increased on average by almost 3% per year, from 1980 to 2005. Rapid growths in developing countries are expected to lead to a similar increase in global GDP, until 2030, resulting in higher demands for energy from all sources (ExxonMobil, 2008).

In addition to economic growth, population growth is also a fundamental driver of overall energy demand. Nevertheless, the types of energy used to meet the specific needs of the different countries depend on additional factors such as availability of supply, income levels and public policies. In 2005, global primary energy consumption was approximately 230 million barrels per day of oil equivalent (MBDOE) of fossil fuels (oil, natural gas and coal) and other non-fossil energy such as nuclear power and renewable sources. By 2030, the total energy demand of the world is expected to increase by 40%, even after assuming significant improvements in energy production and distribution efficiency (ExxonMobil, 2008).

British Petroleum (2008) has carried out a very comprehensive analysis of the consumption of primary energy in different regions of the world. According to this analysis, the consumption of primary energy (commercially traded fuels) at the global level increased by 2.4% in 2007. The IEA estimates that world primary energy demands are likely to increase by more than half between 2005 and 2030, at an average annual rate of 1.8%. Those developing countries whose economies and populations are growing the fastest are expected to account for 74% of the increase in global primary energy use, with China and India accounting for 45% (IEA, 2008). Global demand will be met by a variety of energy mixes. Nevertheless, fossil fuels will continue to provide close to 80% of global energy requirements up to 2030, with oil and gas accounting for almost 60%. Renewable energy still remains a small percentage of total global energy use, in spite of its rapid growth from 2007 (British Petroleum, 2008).

Hydropower is not a major option for future energy production in most developed countries. This is because most of the economic sites have already been developed, or are in the process of development, and because of the strong opposition from social and environmental activists. In 2007, growth in global hydropower generation was 1.7%, slightly below its historic average.

Table 3 shows the countries with the greatest annual hydroelectric energy production and installed capacities. New capacities in China and Brazil, and improved rainfall in Canada and northern Europe, offset drought conditions in the United States and southern Europe (British Petroleum, 2008).

The European Small Hydropower Industry Association has forecasted that small hydro will grow faster than large hydro, even though there is still a large dominance of large hydro in the world. Regarding large hydro, the European Union Climate Change Committee (CCC) is looking to harmonise the approval process for large hydro projects among member states and therefore regulate their sustainability and ensure that environmental aspects are taken into consideration by following international criteria. The objective is to avoid differences in transaction costs and achieve clarity and legal certainty in the carbon trading market so that projects from any member state receive equal treatment (De Brauw et al., 2008). While this is a welcome step towards making carbon credits available from large hydro projects, it is not abiding. Each EU member state has discretionary power when assessing the admissibility of project-based credits.

Table 3 Hydroelectric production and installed capacity in 2006: top eight countries

Country	Annual hydroelectric energy production (terawatt hours, TWh)	Installed capacity (GW)
China	416.7	128.57
Canada	350.3	68.974
Brazil	349.9	69.080
USA	291.2	79.511
Russia	157.1	45.000
Norway	119.8	27.528
India	112.4	33.600
Japan	95.0	27.229

Source: British Petroleum (2008)

Thermoelectric Generation

Water availability is a regional and national concern for meeting future power generation needs. In addition to human consumption, irrigation and industrial development, environment has become an important competitor for the use of water resources, which in turn has an impact on water available for other uses including power generation. Finding viable locations for thermoelectric power plants will become increasingly challenging with time because of tradeoffs needed between energy, environment and water security, as well as land use, economic, social and political considerations.

In the United States, thermoelectric generation represents the largest percentage of electricity production, with coal-based power plants accounting for about half of the electric supply at the national level. The National Energy Technology Laboratory (NETL) of the Department of Energy of the United States has been conducting research on water conservation and management strategies to develop practical solutions to conserve water resources, minimise impacts on water quality and provide environmentally sound solutions for increasing national energy security through domestic resources, for thermoelectric power plants (Feeley, 2004; NETL, 2006). Given that the thermoelectric-generating capacity in the United States is expected to increase by 18% between 2005 and 2030 (EIA, 2008), reliable availability of freshwater in terms of quantity and quality has to be an important consideration.

A fundamental approach for the future has to be to make existing and future thermoelectric power-generating plants more efficient in terms of their water requirements. In the United States, water withdrawals for this sector increased from 492 billion litres per day in 1995 to 553 billion litres per day in 2005. In the future, a determined attempt has to be made to develop advanced technologies that enable the use of alternate sources of cooling water and reduce evaporative and/or drift losses from existing wet cooling towers in order to achieve significant savings in freshwater withdrawal. In the United States, in spite of technological and management advances, thermoelectric power generation will remain a major water consumer for the foreseeable future. Depending on the advances made in technology and management practices, water withdrawal from the sector is likely to range between 424 and 583 billion litres per day by 2030. Since the national water consumption in the country is expected to increase in the future, this will result in increasing competition for water resources (Feeley et al., 2008; NETL, 2008).

The increasing importance of compromises and tradeoffs between various sectors can best be illustrated by the following examples from the United States (NETL, 2008):

> Power Generation Facility Siting Concern about water supply, expressed by state regulators, local decision-makers, and the general public, is already impacting power projects across the United States. For example, in March 2006, an Idaho state House committee unanimously approved a two year moratorium on construction of coal-fired power plants in the state based on environmental and water supply concerns. Arizona recently rejected permitting for a proposed power plant because of concerns about how much water it would withdraw from a local aquifer. In early 2005, Governor Mike Rounds of South Dakota called for a summit to discuss drought-induced low flows on the Missouri River and the impacts on

irrigation, drinking-water systems, and power plants. A coal-fired power plant to be built in Wisconsin on Lake Michigan has been under attack from environmental groups because of potential effects of the facility's cooling-water-intake structures on aquatic life. In February 2006, Diné Power Authority reached an agreement with the Navajo Nation to pay $1,000 per acre foot and a guaranteed minimum total of $3 million for water for its proposed Desert Rock Energy Project. In an article discussing a 1,200 MW proposed plant in Nevada, opposition to the plant stated, "There's no way Washoe County has the luxury anymore to have a fossil-fuel plant site in the county with the water issues we now have. It's too important for the county's economic health to allow water to be blown up in the air in a cooling tower."

Biofuels

Increasing biofuels production can have impact on overall agricultural production, land-use patterns, water use and water quality, depending on which crops are grown, and where, and on the agricultural practices and technology used. Biofuels have enjoyed enormous support during the past years, mainly in the United States (corn based) and Brazil (sugarcane based). In contrast, biodiesel production dominates in Europe, with an important contribution from Germany. The key drivers for the growing interest in biofuels have been its direct substitution for fossil fuel and high subsidies provided to the farmers. However, this sector may face serious constraints to growth in the future due to potential uncertainties such as long-term supply of biofuels feedstock; land, water and environmental constraints; levels of agricultural commodity and oil prices; and the extent of subsidies available from the various governments. Growing biofuel demands in recent times have provided farmers with a greater economic incentive to grow crops for biofuel production, which has contributed to reduced food production and higher food prices.

For present and future use of biofuels, and thus water needed for its production, there are important factors that have to be considered such as subsidies, performance and technology-related targets (e.g. renewable portfolio and renewable fuel standards), tax credits, cap and trade frameworks, carbon taxes, loans and grants (Goldman Sachs, 2007). In the United States, subsidy policies on biofuels have focused primarily on energy security and secondarily on providing support to farmers as part of the overall farm policy. Since biofuels production has been considered to be an important component of the US energy portfolio, its production has been encouraged through the Energy Act of 2005, continuation of the ethanol subsidy at $0.135/L ($0.51 per gallon) and direct payment to farmers for corn and soybeans through the Farm Bill. The US Department of Energy has estimated that biofuels, ethanol and biodiesel from all feedstocks will be able to provide 30% of transportation fuel by 2030 (National Research Council of the National Academies, 2008).

Over the last decade, the US government has implemented a range of both supply- and demand-side incentives intended to increase biofuel production. The supply-side incentives have included grants, loans and loan guarantees as well as the federal Volumetric Ethanol Excise Tax Credit, which provides ethanol blenders/ retailers with a $0.135/L ($0.51 per gallon) of tax credit, and a tax credit of up to 30% of the cost of alternative refuelling property, up to $30,000 for business property. Biodiesels receive a credit of $1 per gallon. The US Department of Agriculture manages a programme that provides grants, loans and loan guarantees to ranchers and rural small businesses for the development of alternative energy projects such as the construction of biofuel plants. The US Department of Energy has allocated a programme that allots $2 billion per year to biofuel loan guarantees. The main demand-side incentive is the Renewable Fuel Standard, which requires that increasing amounts of biofuels be blended with gasoline, from 15 billion litres in 2006 to 28 billion litres in 2012 (Goldman Sachs, 2007).

While the main biofuel used in the United States is now ethanol derived from corn kernels, it is expected that ethanol from cellulosic plant sources (corn stalks and wheat straw, native grasses and forest trimmings) may be used commercially within the next decade because of technological breakthroughs. The National Research Council of the National Academies (2008) considers that during the next 5–9 years, increased agricultural production of biofuels is unlikely to have significant impacts on water quantity at the regional and local levels, except in places where water availability is already a constraint. In terms of water quality impacts, these are expected to be in terms of pollution due to increasing use of agricultural chemicals and to soil erosion. These impacts can be partially mitigated by using appropriate agricultural practices and technologies that are already available.

Biofuel production will require more and more water in the future if this sub-sector expands, as some expect at present. As the use of agricultural chemicals like pesticides and fertilisers increases to improve the yields of the biofuel crops, water bodies around such production systems may witness higher levels of nonpoint pollution, which has proved to be significantly more difficult to control compared to point sources of pollution, even in developed countries. Accordingly, the production and processing of the biofuel crops are likely to bring with them attendant water quantity and quality implications. As long as these implications are clearly thought through in terms of their social, economic and environmental impacts, and appropriate remedial measures are implemented as and when required, the problems may be manageable. However, as of now, virtually no country has carefully analysed the water, land, environmental and social implications of increasing biofuel production and then made appropriate policy decisions.

It should also be noted that, just as the energy sector requires large quantities of water, the water sector is an equally important user of energy for its own operation. Energy requirements for pumping are already very significant in nearly all countries. As the number of water and wastewater treatment plants increase exponentially in the coming years, especially in the developing world, the energy needed for their proper operation and maintenance will also increase. Thus, the water and energy sectors will be even more closely interlinked in the future than they are at present, which will require increasing coordination of policies related to the management of these two sectors, requiring many tradeoffs and compromises.

Water and Environment

The implications of environmental management policies on water development and management have received increasing attention during the past four decades, when it was realised that water and environment policies affect each other in many significant ways, sometimes positively, but at other times adversely.

During the early part of the global environmental movement, the primary focus was on how to stop all types of pollution. For example, during the United Nations Conference on the Human Environment, held in Stockholm in 1972, the main water-related environmental concerns considered were preventing water pollution and impacts of acid rain on forests and lakes. Later on, a backlash developed from certain sectors of society on all types of large infrastructure development projects. This was especially relevant for the construction of large dams and irrigation projects. In this "small is beautiful" era, all large development projects attracted considerable criticisms because of their potential adverse impacts, some of which were justified, but some of which were also fictional.

During the 1980s and 1990s, large water development projects from all over the world came under considerable criticisms from social and environmental activists and many nongovernmental organisations (NGOs). This movement probably reached its peak with controversies associated with the construction of some large dams like Sardar Sarovar and Tehri dams in India, Arun II dam in Nepal and the Nagara Barrage (to prevent saltwater intrusion) in Japan. These controversies had both positive and negative impacts on future water management practices, some of which will be briefly discussed next.

On the positive side, many social and environmental considerations that had not been adequately addressed earlier started to receive accelerated attention. Environmental and social impact analyses became the norm rather than exception, and issues like involuntary resettlements and adverse environmental and ecosystem impacts due to large infrastructure development projects became important considerations. Indeed, the pressure from certain sectors of society was such that not only did the earlier shortcomings receive considerable attention, but also planners and policy makers were forced to respond to them promptly and adequately. Consequently, many undesirable aspects of development activities were properly considered and often appropriate ameliorative actions were taken. This probably would not have happened within a short timeframe of only two decades unless concerted opposition had materialised on the construction of such large development projects.

On the negative side, this opposition to large projects ensured that some water development projects that should have been constructed for poverty reduction, employment generation and raising the living standards of the people were seriously delayed. This opposition further contributed to reduction in funding support to water projects from international sources because of the controversies surrounding them, which consistently received adverse national and international media attention. For some unexplained reasons, large water development projects attracted more controversy than other types of development activities. Donors became very reluctant to fund water projects, irrespective of their overall needs and benefits during the 1980s and 1990s.

The discussions started to become more balanced during the post-2000 period. It is now increasingly realised that water development projects must receive priority attention in all developing countries, not only to satisfy domestic and industrial needs, but also to improve food and energy security of the nations. Equally, however, these structures need to be planned and managed in such a way that they are technically feasible, economically efficient, socially acceptable and environmentally friendly. In future, many tradeoffs and compromises will have to be made between all these and other associated requirements since they are not mutually exclusive. This will be a major challenge.

As societal perceptions have changed and the knowledge base to plan and manage water infrastructure has increased significantly, it is now possible to improve the earlier planning and management practices significantly by concurrently maximising the positive economic, social and environmental impacts, minimising the negative impacts and ensuring that the people who are likely to pay the costs of the projects (e.g. those who have to be resettled involuntarily) are explicitly made the direct beneficiaries of the projects. With this changing mindset and better understanding and appreciation of environment–development links, it is likely that the overall discussion of water development and environmental issues will become more objective and less polarising in the future.

While the adverse social and environmental impacts (real or imaginary) of large water developments have received considerable attention from the media and policy makers, another environmental issue has undeservedly received somewhat benign neglect. This is increasing water contamination from point and nonpoint sources because of accelerating domestic, industrial and agricultural activities all over the world. There is no question that water quality management must receive accelerated attention in the future.

Similarly, provision of clean water supplies has received considerable attention from policy makers in developing countries and from national and international institutions, but commensurate interest in wastewater collection, treatment and disposal has often been conspicuous by its relative absence. Regrettably, there are only limited signs that this attitude is starting to change. In the coming decades, proper management of wastewater from domestic and industrial sources must receive as much attention as water supply, if not more, because of the inadequate consideration in the past.

Increasing water pollution from all sources is a major issue for nearly all developing countries, and control of nonpoint sources of pollution is an urgent requirement for all developed and developing countries. Unless the present perceptions and attitudes change radically during the coming decades, water quality management will become a very critical issue for the future. This is because, at the domestic level, nearly all the water that enters any household is eventually discharged as wastewater. Accordingly, introduction of new sources of water to an area, without adequate provision for treating the wastewater, which the introduced water will invariably become, will be only storing up problems for the future.

Even in the many urban centres of the developing world where wastewater is collected through sewer systems, it is often discharged to freshwater bodies, land or oceans with only limited, or even no, treatment. This means that the problem of wastewater contamination is not being solved: it is simply being transferred from one location to another. The underlying implicit philosophy has been somewhat akin to "out of sight, out of mind".

Compared to domestic wastewater disposal, the situation is becoming even more serious and complex with industrial wastewater discharges, which, for the most part, receive inadequate treatment in nearly all developing countries. At present, few urban centres in developing countries have functional secondary and tertiary industrial wastewater treatment plants. Primary wastewater treatment plants are often nonfunctional for significant periods of time because of poor design, inadequate management, lack of political interest and funding, public apathy and many other associated causes. Even when these plants do function, most operate below their design efficiencies. Since domestic wastes are primarily organic, they degrade over a limited time. However, the situation is more complex and serious for industrial wastes, which contain significant amounts of substances that may be toxic to human beings and ecosystems and which are not easily biodegradable.

With fast industrial and urban growth, proper wastewater management is rapidly becoming a serious social, economic and human health issue in nearly all developing countries. In addition, as the nearby surface water and groundwater sources for urban centres are becoming increasingly contaminated with domestic and industrial wastes, these bodies will require higher levels of treatment before they can be safely used as sources for drinking water downstream. The treatment processes needed to decontaminate polluted sources are likely to become increasingly sophisticated and expensive in the coming decades, which may not be an attractive or feasible alternative for many urban areas because of economic and technology management constraints.

The previous discussion refers only to point sources of contamination from domestic and industrial users: nonpoint sources are at present almost totally neglected in developing countries and inadequately managed in developed countries. The use of agricultural chemicals in many developing countries is still somewhat limited. Accordingly, nonpoint sources of pollution are still not as serious as point sources. However, as there is increasing emphasis on increasing crop production per unit area to enhance both farmers' incomes and food security, more and more agricultural chemicals are likely to be used in the future by farmers in developing countries. This will further aggravate the water quality conditions, because control and management of nonpoint sources of pollution are very complex and difficult tasks under the best of the circumstances. Even the most developed countries, like those belonging to the European Union as well as Japan and the United States, have found it very difficult to manage nonpoint sources of pollution. Agricultural chemicals, which are extensively used in developed countries, leach into the rivers which ultimately carry them to the estuaries, increasing concentrations of agrochemicals in the estuaries and the oceans around them. Consequently, worldwide, more dead zones in the estuaries of major rivers like the Mississippi are being observed, and many of these dead zones are expanding with time. In the coming years, reduction or even elimination of the dead zones through effective management of nonpoint sources of pollution will be an important consideration.

In a macro-global sense, a major challenge facing developing countries is how quickly and efficiently current wastewater management practices and processes can be substantially improved. Considering the cost of construction and efficient operation of wastewater management plants, and the number of trained and experienced personnel needed to manage them, ranging from managers to plant operators and technicians, who are mostly not available at present, resolution of this problem in the foreseeable future will be a most difficult task.

Another macro issue in the water and environment area of the future is likely to stem from the increasing acceptance of the concept of environmental flows. Many countries have now accepted, or are in the process of accepting, that the environment is a legitimate user of water. This means that certain quantities of river flows should be earmarked for environmental and ecosystem uses.

It is highly likely that in the foreseeable future there will be increasing acceptance of this concept in the world as a whole. This will present two types of problems: one conceptual and the other practical. At the conceptual level, considerable additional research needs to be conducted on how environmental flows of rivers can be reliably estimated for both perennial and ephemeral rivers for various regions with different climatic regimes, physical and ecosystem conditions, social and economic situation of the people and many other associated conditions. The development of methodologies to reliably estimate environmental flows under differing conditions has to be an important requirement for the future.

At the practical level, available amounts of water in many rivers have already been allocated and, in many cases, over-allocated, especially during dry seasons and drought periods. Under such conditions, new allocations of water to the environment will mean that some of the existing allocations to domestic, industrial and agricultural sectors will have to be reduced, which will be a difficult task because of social, economical and political reasons. In addition, for transboundary rivers, as well as interstate rivers in federal countries like Brazil, India, Pakistan and the United States, this will raise new sets of legal and institutional issues, especially when inter- and intra-country treaties already exist for water allocations to various state parties. Considering it often takes 20 years or more to negotiate new water allocation treaties for transboundary and interstate rivers, implementation of the concept of environmental flows in such water bodies may prove to be a complex, difficult and time-consuming task.

Finally, environmental impacts of natural disasters on water and wastewater infrastructure cannot be ignored. As much as possible, infrastructure has to be designed to withstand floods, earthquakes and other natural disasters like tsunamis and storm surges. In other words, future water–environment interactions must be viewed through a much broader conceptual framework compared to what is being practiced at present. Appropriate implementable frameworks thus need to be developed.

Technological Advances

Like climate change, technological developments are likely to introduce another set of uncertainties in water management practices and processes. However, unlike climate change, technological developments are much more likely to bring positive surprises in numerous aspects of water development and management.

The information and communication revolutions have had radical impacts on water. Management and analysis of water-related data have become a far simpler and more economic and efficient process than ever before in human history. Information storage, retrieval and exchange have improved exponentially in recent years. South–South knowledge transfer, which was in its infancy some 25 years ago, has now come of age because of tremendous advances in information management and exponentially declining costs. In future, such advances are likely to progress even further. These developments will unquestionably have significant implications for the water sector, including more effective ways to conduct sustainable communications and interactions with the various stakeholders on a regular basis.

Another area that will have a major impact on water-use patterns will be biotechnological advances. These advances will help in the development of pest- and drought-resistant crops, as well as crops that can be grown in marginal-quality water, like saline water. The net impacts of these likely developments may be that more crops can be grown with lower quantities of water and also with the use of marginal-quality water.

Biotechnology is likely to help in many other ways. For example, a new variety of rice under field trial at present can survive for 3–4 weeks under flood water. Every year, hundreds of thousands of tons of rice crops are lost due to prolonged submergence under flood water. These new varieties of rice crops will be able to withstand most flooding.

Similarly, biotechnology is making rapid advances in wastewater treatment. It is highly likely that there will be further very substantial improvements and breakthroughs in these areas during the coming decades. These could have profound effects on water quality management, which is now a very serious problem nearly all over the developing world.

Another area where technological developments have made remarkable progress during the past decade is in desalination. Reduction in desalination costs has made this an important alternative for increasing water availability for both domestic and industrial sectors. By using the new generation of membranes and improved management practices, seawater desalination costs have fallen by almost a factor of three during the past decade. At the current cost of producing desalinated water (around $0.45–0.60 per m^3) through reverse osmosis, the technique has become cost-effective for many cities where water availability is a constraint. The cost of treating brackish water has become even lower: $0.20–0.35 per m^3, depending on its salt content. The technological and management breakthroughs achieved are making desalination a viable alternative for solving water quantity and quality problems for domestic and industrial uses, especially for coastal areas. However, there are many other factors like energy availability, technology management and environmental considerations (especially disposal of brine), which need to be carefully assessed before desalination practices can be successfully and extensively used on a sustainable basis in any country. Desalination-related issues have been discussed in greater detail in a later chapter of this book.

The water profession, in general, has not fully appreciated the potential applications of technological advances which are likely to change water use and demand patterns very significantly. An important reason for this non-appreciation is that water managers have very little, if any, regular contact with professionals from other sectors like biotechnology, information and communication technology and desalination where these developments are taking place and which are likely to influence how water will be managed in the future. Extensive interactions with such professions will be an essential requirement for the future.

However, even when the new technologies become available and cost-effective, national capacities to manage them properly need to be developed. Capacity building for managing water resources in the coming years, in spite of considerable rhetoric, is still not receiving enough attention in most countries. It should be realised under rapidly changing global conditions that tomorrow's water problems can no longer be identified, let alone solved, with today's knowledge and yesterday's experience. A whole new mindset will be needed to identify and solve future water-related problems, which will require substantial attention and additional investments to capacity building.

All the existing and likely future trends indicate that there will be extensive technological advances that will significantly contribute to the solutions of many of the future water problems of the world. Equally, there will be new sets of constraints for timely technology adoption that will have to be overcome. Both the opportunities and constraints may differ from country to country and even within a country, and technological solutions may be location specific. Those countries that will make a determined attempt to adopt emerging technologies to solve their water-related problems will make remarkable progress in assuring water security. In principle, availability of appropriate quantity and quality of water should not be a constraint for them to improve human welfare.

Planning for Uncertainties and Unexpected Developments

Uncertainties associated with efficient water management have increased substantially over the past decade due to factors over which the water sector has control (such as regulations, technologies and demand patterns) as well as those over which the water professionals at best may have only limited control. Water management in the future will thus have to be carried out under increasingly uncertain conditions.

One example of such uncertainties is climate change, a main global concern at present. Irrespective of all its uncertainties, climatic changes and/or fluctuations harbour the risk of a growing number of extreme weather conditions and catastrophes whose social and economic impacts could be very significant and likely to increase with time. In most countries, natural disasters disrupt the normal process of economic development. Developing countries already suffering from resource and capacity constraints are often forced to divert their limited resources from ongoing development activities to immediate relief-and-rescue operations because of disasters, which can sometimes set back their development plans by as much as a decade (Grabs et al., 2007). During the last 50 years, climate-related losses have increased very significantly, with the trend becoming more marked since about the mid-1980s. For example, mean annual economic losses caused by major climatic-related catastrophes increased from $12 billion to $40 billion during the 1990s (Munich Re Foundation, 2007). The drought of the early 1990s in Zimbabwe was associated with an 11% decline in GDP; the recent floods in Mozambique led to a 23% reduction in GDP; and the drought of 2000 in Brazil cut the projected economic growth by half (Lenton et al., 2005).

Another uncertainty stems from economic growth and the impacts of massive energy-related developments of newly emerging economies such as BRICs, as they become major economic and trade partners, competitors, resource users and polluters. These are on a scale that was previously unthinkable. Their implications for the water sector are still only partially appreciated and understood, but it is clear that the effects have transcended their national boundaries. The increasing social and environmental pressures are being felt mainly by developing countries at present because they are less equipped than developed countries to make the necessary financial and institutional adjustments within limited timeframes. Even for developed countries, these changes have largely outpaced the benefits of any efficiency gains that have been witnessed in recent years. New and innovative policy actions are thus urgently needed for more efficient management of natural resources, including water, for both the developed and developing world (OECD, 2008).

At a local level, institutions will be forced to respond to social and economic uncertainties. For example, water utilities will have to increase their efficiencies to respond to the needs of more active and demanding consumers than ever before and also to the wishes of their political masters, who in turn will be reacting to societal demands for better services, perhaps with a time lag. Planning for the provision of good and acceptable water services will include consideration of many different types of uncertainties, among which will be population growth and structure (which may translate into additional resource requirements and different contaminant loads), changing political landscape (including increasingly important roles of NGOs and other civil society organisations in the formulation and implementation of public policy decisions), regulations in a highly politicised environment, nature of the workforce (including retirement of skilled and experienced workers), increasing and more effective use of existing and new technology, utilisation of new sources of water through desalination or reuse of treated wastewater, use of marginal-quality water, financial management including using water tariffs to finance and manage infrastructure, and managing cost of energy, which will remain a large component of the cost of production and distribution of water and disposal of wastewater (Jeans et al., 2005).

In terms of public participation, the involvement of multiple actors with diverse interests in water management, along with the increasing importance of issues such as responsibility, accountability, transparency, equity, fairness and corruption, will further increase the complexity of water management from the global to the very local levels. To these uncertainties have to be added a continuing overall deficit in terms of issues like good governance, efficient institutions, adequate and timely financial investments and political will (Tortajada, 2008b). Together, these factors are as much a cause of global water imbalances and driving forces as are trends in population growth, urbanisation and economic development.

Even though water institutions all over the world are facing increasingly uncertain conditions, solutions still remain mostly traditional. So far, advances have been mostly incremental. Examples include many past strategies to water quantity and quality concerns which have been narrowly focused without appropriate consideration of their financial viabilities on a long-term basis; urgently needed water pricing reforms, which are often missing; absence of effective public–private partnership; regulatory frameworks that do not encourage efficiency or innovation; and infrastructural developments that continue to neglect poverty alleviation. In addition, there are continuing shortcomings with technology absorption and adoption processes, capacity building and forward-looking education, training and research programmes.

The global water community must engage more actively on issues related to knowledge generation and synthesis from different parts of the world, not only from within the water sector, but also from *outside*, since these will have a bearing on water management in the future. This should include consideration of appropriate policy options to solve key water-related problems due to their increasingly complex, and often cross-sectoral, nature, but always within the framework of social and economic development of the country concerned. Different actors need to work together to ensure the formulation and implementation of coherent water management policies by considering the problems of the future, rather than focusing on issues of the past which may no longer be important or relevant. Ensuring improved and sustained communication between a multiplicity of groups having different interests and agendas, dissimilar ethics, values and norms and absence of an overall consensus about the types of goals that are to be pursued will present formidable challenges that will require extraordinary measures of coordination, collaboration and cooperation which simply do not exist in the water sector at present. All these issues will further increase the uncertainties associated with water management processes in the future.

One of the most important requirements for the future for both developed and developing countries will be accelerated priority attention to water quality management. Each year, the numbers of people who are affected at the global level by waterborne diseases are estimated to be in the order of millions, and the related costs are likely to be in the order of billions. However, there are very few studies available at present which can be considered to be definitive because of their methodological and data constraints. The estimates that are available at present are mostly based on simplistic and erroneous assumptions and poor quality data: they are highly likely to be very wide off the mark.

The Third World Centre for Water Management has carried out analyses for Mexico (Marañón-Pimentel, 2009). According to this study, in 2005 alone the costs due to water-borne diseases in the country were estimated to be $260 million, with the highest mortality rates in population between 0 and 4 years and above 65 years. Nevertheless, and in spite of the economic importance of the impacts that waterborne diseases have had in the country for decades, these costs have never been estimated reliably by the various governmental institutions or academia.

Furthermore, widespread mistrust of the quality of tap water that is provided all over Mexico has ensured that all levels of the society are paying high economic and social costs in terms of access to clean drinking water, irrespective of their socioeconomic status. Drinking water for much of the population, even when they have access to tap water, comes mostly from 20-L containers which are sold commercially and whose quality is also questionable (Marañón-Pimentel, 2009).

The perception of the population that the quality of tap water is not suitable for drinking has had an enormous impact on the economy of the country. In 2007, Mexico became the largest per capita consumer of bottled water in the world, with total consumption increasing from 11.6 billion litres in 1999 to 22.33 billion litres. At present, per capita bottled water consumption in Mexico is nearly twice that of the United States, even though its per capita GDP is about one-sixth that of the United States.

It is not only the economic costs of bottled water that the consumers are forced to bear in countries like Mexico, India, China or Brazil because the quality of domestic water supplied leaves much to be desired, but also the social and environmental costs that are quite high. For example, worldwide, it is estimated that 2.7 million tons of plastic are used annually by the bottled water industry. Energy requirements for the production and distribution of the products are high. Environmentally sound disposal of the empty bottles is a problem in nearly all developing countries.

With present knowledge, management practices and technology available, there is absolutely no reason why important urban centres of countries like Mexico, India, China, Brazil or Egypt cannot be provided with clean drinkable water on a 24-hour, 7 days a week basis. Much of the funding needed to provide 24 × 7 water supply is already available but, most unfortunately, such funds are not being efficiently used at present. The main reasons for this unsatisfactory performance of the water utilities is continued poor governance; lack of realisation by the policy makers that water quality is an important issue for health and environmental reasons and that the problems are solvable; and public apathy, extensive political interferences, inappropriate institutional structure and widespread corruption. A major future uncertainty will stem from how long it will take the public in developing countries to demand, and then get, drinkable water supply and acceptable levels of water management. Until the existing public apathy disappears, bureaucrats and politicians will continue to give lip service to the importance of proper urban water management, which will mean that the likely improvements in the future will be mostly incremental. The economic, social and environmental costs will continue to increase, and an apathetic public will continue to accept the poor services offered. The events that will trigger the changes whereby the public will forcefully demand better water-related services, and the timeframes over which such changes are likely to occur in different regions, are very difficult to predict at present.

Final Reflections

Projections of future population and consumption trends indicate that demands for water of appropriate quantity and quality for various uses will be an issue for all regions, in both developed and developing countries. Drivers such as urbanisation, population, industrialisation and economic development and the corresponding rises in the demands for food, energy and environmental security are just a few of the trends that will seriously affect existing water planning, management and allocation processes.

Degradation of natural resources, including water, has been due to continued mismanagement and poor governance. Extensive policy and market failures have received limited corrective actions from the concerned institutions over the past decades. Rapid urbanisation, industrialisation and economic growth have imposed complex demands not only on the environment but also on the human and the institutional abilities to respond to such needs efficiently. The net result has been misuse and over-exploitation of resources in most countries of the world. There is now an urgent need to formulate forward-looking, business-unusual water policies and strategies that can reform and strengthen public institutions, increase public and private sector investments, manage urban and rural environments, encourage use of available and appropriate technologies, seriously consider South–South technology transfer and develop a new generation of capable managers and experts representing various appropriate sectors and disciplines with good communication skills.

Given the previously discussed constraints, trends and drivers, new visions for water futures need to be formulated for national, regional and local levels. While there is likely to be some common elements for the formulation of these futuristic visions, each one of them has to be developed for specific conditions. These visions should be developed for at least 10–20 years, which will be a radical departure from the current practice where the plans are for 4–6 years. Furthermore, in the past, as well as at present, water management policies and plans have been mostly framed narrowly on a sectoral basis, with very limited consideration of future drivers from other sectors that are likely to affect the water sector. There is continuing emphasis on short-term solutions based on past experiences and prevailing bandwagons, which are unlikely to provide the right long-term solutions for the new generation of emerging water problems.

On the basis of the analyses carried out at the Third World Centre for Water Management, it can be said with considerable confidence that, as the twenty-first century progresses, the water profession will face a problem the magnitude and complexity of which no earlier generation has had to face. The profession at present faces two main choices: carry on as before with a business-as-usual and incremental attitude and endow future generations with a legacy of poor water governance and a plethora of partially resolved water-related problems or continue in earnest an accelerated effort to identify, understand and then efficiently manage the likely problems of the future, in order to ensure that the world's water resources are properly managed on a long-term basis for the benefit of the entire humankind.

All the major issues facing the world will become increasingly interrelated. The dynamics of the human future will not be determined by any one single issue but by the interactions between a multitude of issues. Increasing population, urbanisation, globalisation and standard of living will require more food, energy and other efficient raw materials, as well as their management which must be significantly more efficient than ever before. Assuring food, energy and environmental securities will necessitate good water governance on a long-term basis. The common requirements for all practical responses to the solutions must include greater and efficient investments; use of more knowledge, technology and expertise; eschewing of dogmatic and/or solution-in-search-of-a-problem approaches; functional institutions; and intensified cooperation and coordination between sectors as well as within countries and between countries.

The interrelationships between these issues are global in character, and hence they can probably be best understood, and then resolved, within a global framework. While the framework may be global, within this there must be a wide variety of well thought through and coordinated national and local responses. Water problems of the future need to be viewed, analysed and solved within global, regional, national and local frameworks. This will be a radical departure from the existing practice.

The water profession should realise that the world is heterogeneous, with different physical attributes, economic and climatic conditions, cultures and social norms, availability of natural resources, management capacities, and institutional arrangements. The systems of water governance, legal and institutional frameworks and modes of decision-making may often differ from one country to another in very significant ways. Under such diverse conditions, one fundamental question that the water profession must ask is whether any single management paradigm can encompass all countries in the future, as often implicitly assumed at present. Can any single paradigm be equally applicable, both now and in the future, for Asian values, African traditions, Japanese culture, Western civilisation and Islamic customs? Can any general paradigm be equally valid for monsoon and non-monsoon countries, deserts and very humid regions? The answers to these questions are likely to be negative. This means that many of the water management paradigms that are popular at present are unlikely to be very useful in the future.

The world is changing very rapidly, and with it the current water management practices must change as well. The types and nature of future water problems must be carefully anticipated and then objectively analysed in the light of the expected changes. In the final analysis, it is deeds and not rhetoric or dogmatic beliefs that will be most important in solving future water problems. Past or current solutions are unlikely to shed meaningful light on the coming, new, turbulent and uncertain world of water management.

During the next two decades, policy makers will have to juggle regularly with the competing, conflicting and changing needs of water for different purposes and by various stakeholders, as well as to coordinate effectively the increasing needs of concurrently assuring water, energy, food and environmental securities in order to maximise human welfare. Water will be one of the important common threads that will bind all the four concerns. The task facing the water profession will be historically unprecedented and will make proper adequate understanding of all the different pathways of interlinkages between different sectors, users and interests, numerous cause-and-effect interrelationships and a series of implications and consequences that will be very difficult to forecast and analyse and even more difficult to manage under constantly changing conditions.

There is now a revolution taking place in water management, even though most professionals may not be aware of it. In the wake of this revolution, long-held concepts and models of water management are likely to evolve further in an accelerated manner, and some may even disappear completely. Never before in human history has water management faced so many profound changes within such a short period of time, as are likely to be witnessed during the next two to three decades. How to anticipate and manage these expected changes successfully and in a timely manner will be a major challenge that the water profession must meet. In this connection, it will be desirable to heed the advice of the eminent eighteenth century British statesman and philosopher Edmund Burke: "never plan for the future by the past". In the twenty-first century, Burke's advice will be more prophetic than ever.

References

Biswas AK (2006) Water management for major urban centres. In: Varis O, Tortajada C, Lundqvist J, Biswas AK (eds) Water management for large cities. Routledge, Abingdon, pp. 3–17

Biswas AK, Tortajada C (2009) Hydropolitics of Aswan High Dam, water resources development and management book series (Biswas AK, Tortajada C (eds)) Springer, Berlin (forthcoming)

British Petroleum (2008) BP Statistical review of world energy, June 2008. Available at http://www.bp.com/productlanding.do?categoryId=6929&contentId=7044622

California Department of Water Resources, California Water Plan (2005 and 2009) Updates. Available at http://www.waterplan.water.ca.gov

De Brauw, Blackstone and Westbroek (2008) Emission Trading: Proposed EU harmonisation of large hydro criteria. De Brauw Legal Alert, May 2008. Available at http://www.debrauw.com

E & WR (2005) In: CSIS. Global water futures. Addressing our global water future. Center for Strategic and International Studies, Sandia National Laboratories, California

EIA (2008) Annual Energy Outlook 2008 with Projections to 2030, Energy Information Administration. http://www.eia.doe.gov/oiaf/aeo/index.html

ExxonMobil (2008) The Outlook for Energy: A View to 2030, Irving, Texas. Available at http://www.exxonmobil.com/Corporate/energy outlook.aspx

Feeley TJ III (2004) Responding to emerging power plant-water issues, DOE/NETL's R&D Program, Presentation at the American Coal Council 2004 Spring Coal Forum, Dallas, Texas, May 17–19, 2004

Feeley TJ III, Skone TJ, Stiegel GJ Jr, McNemar A, Nemeth M, Schimmoller B, Murphy JT, Mandredo L (2008) Water: a critical resource in the thermoelectric power industry. Energy 33:1–11

Goldman Sachs (2006a) Looking Forward to 2007 and Beyond. CEO Confidential, Global Economics, Issue 2006–10, December 1, 2006

Goldman Sachs (2006b) Building the World: Opportunities in infrastructure. CEO Confidential, Global Economics, Issue 2006–2006, June 15, 2006

Goldman Sachs (2007) Alternative Energy: A Global Survey. Global Markets Institute, Fall 2007. At http://www2.goldmansachs.com/citizenship/global-initiatives/research-andconferences/past-conferences/alternative-energy-docs/global-survey.pdf

Goldman Sachs Global Investment Research, Nuclear Energy Overview (March 2007). In: Goldman sachs, Alternative energy: A global survey, Fall 2007, New York

Grabs W, Tyagi AC, Hyodo M (2007) Integrated flood management. Water Sci Technol 56(4): 97–103

IEA (2007) World Energy Outlook: China and India insights. International Energy Agency and Organisation for Economic Co-operation and Development, Paris

Jeans EG III, Ospina L, Patrick R (2005) The primary trends and their implications for water utilities. J Am Water Resour As 97(7):64–77, July

Lawson RL, Lyman JR, McCarthy ER (2008) A 21st century marshall plan for energy, water and agriculture in developing countries. Policy paper. The Atlantic Council of the United States, Washington, September 2008

Lawson S, Purushothaman R, Heacock D (2005) 60 is the new 55: how the 66 can mitigate the burden of aging. Goldman Sachs Global Economic Paper No. 132, 28 September. Available at http://www2.goldmansachs.com/ideas/demographic-change/60-is-the-new-55-pdf.pdf

Lenton R, Wright AM, Lewis L (2005) Health, dignity and development: what will it take? UN millennium development project task force on water and sanitation, Earthscan, London

Marañón-Pimentel B (2009) Economic and health costs related to the lack of reliable drinking water supply services in Mexico City. Int J Water Resour Dev 25(1):65–80, March

Munich Re Foundation (2007) Report, from knowledge to action, Munich

National Research Council of the National Academies (2008) Water implications of biofuels production in the United States, The National Academy Press,Washington, DC National Academy of Sciences, 2008

NETL (2006) Water & energy. Addressing the critical link between the nation's water resources and reliable and secure energy. US Department of Energy, Office of Fossil Energy, National Energy Technology Laboratory, Albany

NETL (2008) Estimating freshwater needs to meet future thermoelectric generation requirements, 2008 Update. DOE/NETL-400/2008/1339, National Energy Technology Laboratory, Department of Energy, United States of America, Albany

NIC (2008) Global Trends 2025: A Transformed World, National Intelligence Council, Government of the United States, Washington, p. 52

OECD (2007) Environmental Outlook to 2030. Organisation for Economic Co-operation and Development, Paris

OECD (2008) OECD Forum 2008 "Climate change, growth, stability", policy brief, special edition, OECD, Paris

OECD/IEA (2008) Key World Energy Statistics, Organisation for Economic Co-operation and Development and International Energy Agency, Paris

Rangachari R (2006) Bhakra-Nangal project, socio-economic and environmental impacts. Oxford University Press, New Delhi

Tortajada C (2004) South-eastern anatolia project: Impacts of the Ataturk Dam. In: Biswas AK, Unver O, Tortajada C (eds) Water as a focus for regional development. Oxford University Press, Delhi, pp. 190–250

Tortajada C (2008a) Challenges and realities of water management of megacities: the case of Mexico City metropolitan area. J Int Aff (Spring) 61(2):147–166

Tortajada C (2008b) Rethinking water governance. In: Feyen J, Shannon K, Neveille M (eds) Water and urban development paradigms. Towards an integration of engineering, design and management approaches, CRC Press, London, pp. 523–541

United Nations, Department of Economic and Social Affairs, Population Division (2007a) World population prospects: The 2006 Revision, Highlights, Working Paper No. ESA/P/WP.202, New York

United Nations, Department of Economic and Social Affairs, Population Division (2007b) World Population Ageing, New York

Varis O (2009) More urban and more aged: demographic pressures to global water resources by 2050. In: Biswas AK, Tortajada C, Izquierdo R (eds) Water management beyond 2020. Springer, Berlin

Wilson D, Dragusanu R (2008) The Expanding Middle: The Exploding World Middle Class and Falling Global Inequality, Goldman Sachs, Global Economic Paper No. 170, New York, July 7, 2008

More Urban and More Aged: Demographic Pressures to Global Water Resources by 2050

Olli Varis

Introduction

Population growth has largely been considered as the ultimate driver for the sustainability challenges of our planet's development. Whereas this assumption is not totally wrong, the demographic pressure on this planet is a far more complex issue than mere growth and has many important facets. It is, for instance, very interesting to look at populations that do not grow but instead stay constant or even decline.

On the world scale, the population is still growing, but the growth occurs only in certain geographical regions. In other regions, the primary demographic processes are the aging of the population and urbanisation.

The aging problem has typically been associated with low-birth-rate countries in the developed world and many of the former centrally planned economies such as Russia, Ukraine and Belarus, among others. However, this concept is not quite correct:

> Over the next 5 decades, the number of persons aged 60 and over in the developing countries will be 9 times greater than it is today, and the share of elderly persons residing in urban areas will be 16 times greater.

This quotation by Ergüden (2005) gives some dimension to the aging and urbanisation processes in the contemporary developing countries. But, of course, the global wealth distribution might be quite different from the present one after another half a century.

This study aims to provide a global overview of the human demographics of our planet, and in particular, of population aging and urbanisation by the year 2050 and its implications on natural resources, particularly water.

Direct implications to water will be relatively minor in comparison to the indirect ones that come through societies' economic and institutional capacity to provide services and pensions to the elderly population as well as their capability to maintain water services.

O. Varis (✉)
Water Resources Laboratory, Helsinki University of Technology,
PO Box 5200, 02015 Espoo, Finland
e-mail: olli.varis@tkk.fi

Demographic Aging in 1950–2050

The United Nations Population Bureau publishes every 2 years a revision of the global population prospects. This section will summarise the 2006 prospects (United Nations, 2006), with particular focus on urbanisation and aging of the human population.

It is important to note that the classification used by the United Nations Population Division contains some features which are not the most commonly used ones in geographic analyses. Therefore, the specification of these regions is provided here. For more details, see United Nations (2006). The acronyms for these regions used later in this study are also given:

- Europe (E): Includes the European countries, excludes the southern Caucasian states Armenia, Azerbaijan and Georgia. Includes the whole territory of Russian Federation, but excludes Kazakhstan and the other four Central Asian states.
- North America (NA): Does not include Mexico.
- Oceania (O), China (C), Southeast Asia (SEA): As usual.
- East Asia (EA): Excludes China.
- South Asia (SA): Also includes Iran, Afghanistan and the five Central Asian states.
- Middle East and North Africa (M; MENA): Also includes the southern Caucasian states, Cyprus, Turkey and the Sudan.
- Sub-Saharan Africa (SSA): Excludes the Sudan.
- Latin America and the Caribbean (LAC): Includes Mexico.

The number of child births has been stabilising in the past two decades to between 132 and 137 millions per year (Fig. 1) and is projected to start to decline slowly within a decade. In this sense, the world's population expansion is already under relatively good control. The population is, however, still growing at around 75 millions per year. This is because the number of deaths is 75 millions lower than

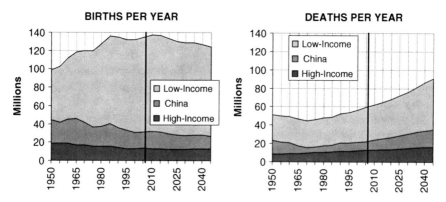

Fig. 1 The evolution of human births and deaths from 1950 to 2050
Source: United Nations (2006).

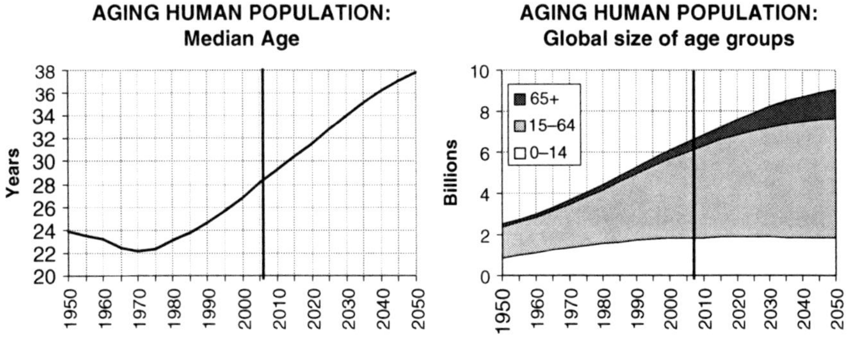

Fig. 2 Aging global population: the evolution of the median age of the human population and global population growth across different age groups (in billions)
Source: United Nations (2006).

the number of births. The number of deaths is growing substantially, which means that the growth rate of the human population is in decline.

The age structure of the world's population is undergoing a rapid change; the median age has climbed from 22 in the 1970s to the present 28 and is projected to reach 38 by 2050 (Fig. 2). This means that the median age grows around 3 months in 1 year. Consequently, the aging of the human population has become one of the major demographic issues and challenges.

The population aged 65 years and over was 131 millions in 1950. It had climbed to 480 millions by 2006 and is expected to increase to 1,465 millions by 2050. This means that the share of senior citizens has grown from 5.2% in 1950 to 7.5%, but expansion is soon to come; by 2050 it is projected to reach 16.1%. This growth becomes more concrete when we look at the figures by region (Table 1).

The following age classification will be used throughout this analysis:

- 0–14: infant population below 15 years of age
- 15–64: working-age population between 15 and 64 years of age
- 65+: population aged 65 years and above

Table 1 Size of the aged population (65+) by region in 1950, 2005 and 2050

	Year	Europe	Other HI	LAC	China	SA	SEA	MENA	SSA
Millions	1950	45	20	6	25	18	7	4	6
	2005	116	77	34	100	79	29	18	23
	2050	180	165	144	329	340	129	92	91
Indexed	1950	1.0	1.0	1.0	1.0	1.0	1.0	1.0	1.0
	2005	2.6	3.8	5.5	4.0	4.3	4.4	4.5	4.1
	2050	4.0	8.2	23.2	13.2	18.4	19.0	22.6	16.0

Source: United Nations (2006).

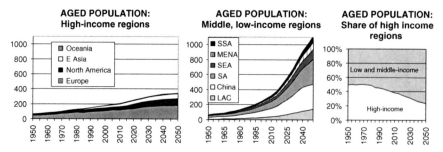

Fig. 3 The growth of the aged population (65+ years; in millions) in different regions of the world and the global share of total aged population in high-income countries
Source: United Nations (2006).

Aging is often associated with wealthy countries, but in fact there are several middle- and low-income regions in which the aged population is also growing rapidly (Table 1, Fig. 3). The global share of high-income countries is decreasing, and around 40% of the world's 65+ population lives in high-income countries. This number will go down to 20% by 2050. Accordingly, aging is very much an issue of the developing world as well.

- Europe's aged population was 45 millions in 1950. It is now 116 millions and is expected to reach 180 millions by 2050.
- In other high-income regions, the development is somewhat faster than in Europe.
- The aged population grows fastest in middle-income regions such as Latin America and the MENA region.
- China has a far slower growth of aged population than other developing regions, but the share of its aged population grows faster than elsewhere else.

A commonly used numerical indicator for aging is the potential support ratio (PSR), which is the ratio of the working-age population (15–64) to aged population (65+). If PSR is 4, it implies that there are four working-age persons for one senior citizen. This analysis also uses its inverse 1/PSR, since it is in some cases more indicative of real growth rates than PSR (Fig. 4). 1/PSR tells what percentage the size of the aged population is if it is compared to the working-age cohort.

In 1950, Europe, North America and Oceania had much lower PSR in comparison to the rest of the world. The 1/PSR of the first two was around 12–13%, while it was 6–8% for the third. For the first two, the ratio has gone up to 15–22% at present, whereas for the third, it is now between 6% and 12%.

Europe has been the region with the highest share of aged population. But East Asia is just passing Europe, and it is interesting to note that East Asia has shown a very rapid growth of the share of aged population, and the growth keeps accelerating.

China's 1/PSR has modest growth and is passing 10%. However, the growth will accelerate dramatically after 2010, and will reach Europe's and East Asia's present

More Urban and More Aged

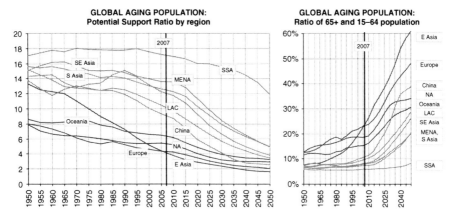

Fig. 4 The potential support ratio (PSR) by region between 1950 and 2050 and the ratio of aged population (65+) and the working-age population (15–64 years). Note that the right figure shows 1/PSR
Source: United Nations (2006).

level by 2030. In fact, China's aging pattern looks very similar to that of the rest of East Asia, but with a delay of 25 years.

Latin America and Southeast Asia have still a relatively small aged population, but they will start to grow rapidly in the coming decade. These regions are expected to reach North America's present aging level by 2030–2040. Middle East and North Africa as well as South Asia are aging somewhat more slowly than Latin America and Southeast Asia. Sub-Saharan Africa's aged population will remain proportionally very low for several decades ahead.

Whereas the growth rate of the aged population (65+) is very high, the very aged population (80+) still grows at a much higher pace (Fig. 5, Table 2). In 1950, 1% of the population of the more developed regions had reached the age of 80; now their

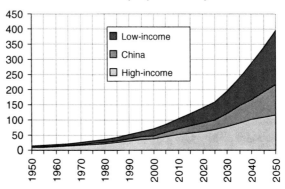

Fig. 5 The number of very aged population (80+; in millions) from 1950 to 2050
Source: United Nations (2006).

Table 2 Size of very aged population (80+) in high-income countries, in China and in other countries (marked here as low-income) in 1950, 2005 and 2050

	Year	High income	China	Other	World
Millions	1950	8.5	1.6	3.7	13.8
	2005	44.5	15.0	27.2	86.6
	2050	115.8	101.8	176.6	394.2
Indexed	1950	1.0	1.0	1.0	1.0
	2005	5.2	9.6	7.4	6.3
	2050	13.6	65.1	47.8	28.6

Source: United Nations (2006).

share is 3.7%, and is expected to reach 9.7% by the year 2050. In absolute figures the growth is phenomenal: from 8.5 millions in 1950 to 44.5 millions in 2005 and up to 116 millions by 2050.

In less-developed regions the growth is still more impressive. In 1950 those regions had 5.3 million people over 80 years old, which comprised 0.3% of their total population. These figures grew to 42 millions (0.7%) by 2005 and are expected to multiply to 278 millions (3.6%) by 2050.

China deserves special mention. Its very aged population grew almost 10-fold between 1950 and 2005 and is expected to grow at nearly the same pace to 2050. Within 100 years, the Chinese population that has reached 80 years of age is expected to grow from 1.6 millions to 102 millions.

Urbanisation in 1950–2050

Between 2006 and 2010, the world's population has been estimated to grow by 75.6 million people per year; 89% of this growth is in urban areas (United Nations 2006). Almost all of this takes place in developing countries, though mainly it is due to migration; fertility rates are far lower in urban areas than in rural ones.

In Africa and Asia, the proportion of urban to rural population is around one to three, while in all the other continents it is more than two to three. Urbanisation is therefore expected to be most rapid in these continents (Fig. 6).

In Sub-Saharan Africa, the urban population is projected to grow faster than in any other region; it grew 13-fold from 1950 to 2005 and will still grow 2.3-fold by 2030 (Figs. 7 and 8, Table 3). In all other regions of the world except Europe, the pace of urbanisation has been astonishingly similar; from 1950 to 2005 it has ranged from 6- to 8.8-fold. The projected rates in these regions, though, differ considerably. In China and Southeast Asia, the growth will continue faster than in other regions. In the MENA region and South Asia, the growth will also be very fast. The other regions will experience more modest urban population growth.

Rural population is not growing much, and in fact it is projected to start to decline within one or two decades. Rural population will be growing in sub-Saharan Africa,

Fig. 6 The growth of rural and urban population from 1950 to 2030
Source: United Nations (2006)

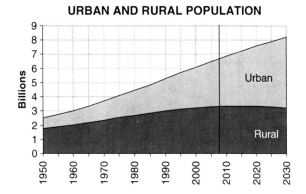

in South Asia and very modestly in the MENA region. Elsewhere, it will decline and in some regions such as China and Europe the decline will be quite considerable. China still has almost 800 million rural residents, but this number will go down to 570 millions by 2050.

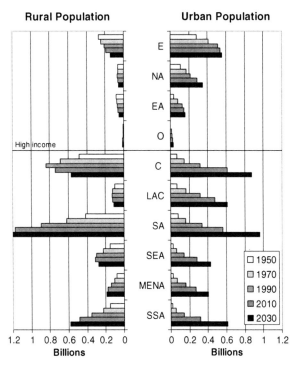

Fig. 7 The growth of rural and urban population from 1950 to 2030 by region
Source: United Nations (2006).

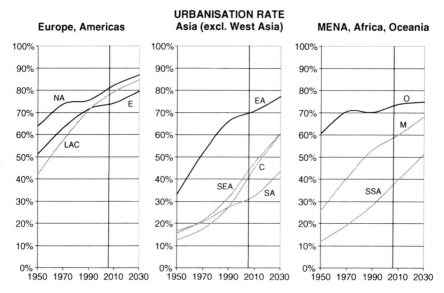

Fig. 8 Percentage of population living in urban areas
Source: United Nations (2006).

Table 3 Size of urban population by region in 1950, 2005 and 2030

	Year	Europe	Other HI	LAC	China	SA	SEA	MENA	SSA
Millions	1950	280	156	70	70	82	27	27	20
	2005	534	970	436	533	495	243	237	264
	2050	555	1416	611	875	961	426	401	612
Indexed	1950	1.0	1.0	1.0	1.0	1.0	1.0	1.0	1.0
	2005	1.9	6.2	6.2	7.7	6.0	8.8	8.8	13.1
	2050	2.0	9.1	8.7	12.6	11.7	15.5	14.9	30.3

Source: United Nations (2006).

Urbanisation, Aging and Development

The world's regions differ fundamentally in their capacity to face the impending aging- and urbanisation-related challenges. It is interesting to study some key development indicators for the world's regions and relate them to urbanisation and aging. Comparison of the urbanisation and aging development in the past three decades (Fig. 9, left) indicates that the MENA region and Latin America have a relatively young population and high share of population in urban areas if compared to other regions. China, and to a lesser degree South Asia and Southeast Asia, are more aged and particularly little urbanised. Sub-Saharan Africa is the youngest population with a still very low level of urbanisation.

China among the less developed regions and Europe among the more developed ones differ from the others when plotting aging against gross national per capita

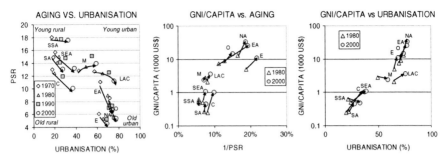

Fig. 9 Urbanisation, aging and economy: developments by region in the period 1980–2000.
PSR = potential support ratio, GNI = gross national income
Sources: World Bank (2003) and United Nations (2006).

income (Fig. 9, middle). They both have clearly a more aged population than the other regions with the same income level. Europe has far more aged people than others, and its economic level is below others. It must be understood here that Europe is a very inhomogeneous region, and taking for instance the European Union separately from the rest of the region would reveal some of the large diversity of the region.

The urbanisation level of the world's macroregions seems to have a close connection to their gross national income (Fig. 9, right). The World Bank (2003) statistics for 1980–2000 reveal that most regions have been able to improve their economies roughly in pace with urbanisation. There are some exceptions, however. In Africa and the MENA region, economies have declined, whereas urbanisation has been rapid. In Latin America and the Caribbean, economic development has been unable to keep pace with urbanisation. In these regions, the social and economic problems that result from urbanisation can be expected to be more severe than elsewhere in the world.

The pace of economic development and the level of education are important factors. Southeast Asia and China have strong economic development and illiteracy rates at the same level as those in Latin America and the Caribbean (Fig. 10). However, while Southeast Asia and China had only 40% of the population in urban areas by 2000, some 80% of Latin Americans were living in cities by then. In the MENA region, literacy levels are far lower than in either of those regions, but urbanisation is relatively high at more than 60%.

Figure 10 indicates the two different paths to urbanisation, one in which people learn to write before massive urbanisation takes place, and vice versa. Obviously, regions that take the former path will have a smoother way ahead in coping with the challenges of urbanisation. Economic development is strongly dependent on levels of education, and there is no way to prosperity if people are denied learning. It will be interesting to see which of the two ways South Asia and sub-Saharan Africa take in the coming years and decades. Hopefully, they will follow the former path; otherwise their problems are expected to grow colossally since neither of them has as enormous external earnings as the MENA region has from oil.

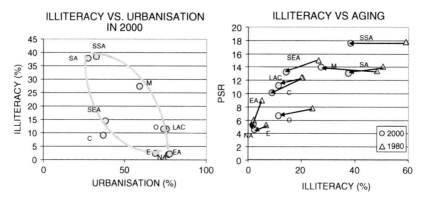

Fig. 10 Urbanisation, aging and illiteracy: developments by region in the period 1980–2000
Sources: World Bank (2003) and United Nations (2006).

Facets of Water for Human Societies

Water has many quite fundamental and quite different facets and functions in human societies. Therefore, the assessment of the effect of a demographic phenomenon such as urbanisation and aging on water is a task with very many aspects.

Water is intertwined in the everyday life of human beings in countless ways. The importance of water as a driver for health, food security and quality of life and as a pillar for economic development is unique. As water affects human lives, humankind also affects the hydrological cycle of this planet, in all dimensions from the very local to the global scale. The production of 1 kg of grain consumes 1000–4000 L water. Food production—although not enough for all—already accounts for 90% of water use in developing countries. Hydropower production by damming rivers evokes grand emotions, yet sustainable energy production is among the cornerstones of economic development. The damage caused by floods and droughts is escalating. The human impact on ecosystems is catastrophic in immeasurable ways. Water is largely a political good since a bulk of humankind lives in river basins shared by two or more nations.

Water is the backbone of the economy in very many countries of the world. Water resources management provides the foundation for the agricultural sector, much of the energy sector, an important part of urban infrastructure, health care and many other functions of the society. Economic growth is desperately needed for poverty reduction, but growth alone is not sufficient. The well-being must reach the poor, otherwise the growth only polarises the economies. Water's role is very important in this complex interplay. Besides being fundamental to many economic sectors, water is also a key to meeting many of the basic needs that are in turn instrumental in poverty reduction:

- *Water*: The more important economically, the poorer the nation is.
- *Environmental threats*: By far the most detrimental environmental catastrophes are floods and droughts. Water is the main carrier of environmental pollutants,

inadequate sanitation often being a major cause for pollutants. It is also the major agent in global erosion, desertification, biodiversity decline and climate change problems.
- *Traditional societies and the traditional sector*: Their economy is tied in with nature and very closely related to the water cycle. Development of water management and sanitation requires culturally tailor-made approaches.
- *Housing and informal sector*: Water and sanitation are key constraints for decent housing and livelihood as well as for the rapidly growing informal sector. The challenges are soaring particularly in urban conditions.
- *Agriculture*: Accounts for 70% of all water use by humankind. In most developing countries, the share is over 90%. Water, nature, infrastructure, technology and so on are the backbone of the economy.
- *Industry*: In large parts of the developing world (China, Southeast Asia, South Asia), industry is developing more rapidly than ever before. Many industrial sectors rely on water. Pollution challenge is enormous and so is the growing thirst for energy.
- *Energy*: The Johannesburg Plan of Implementation defined the increase in the share of renewable energy sources as the primary goal of the energy sector. It is fundamental to understand that 96% of the contemporary renewable energy production comes from either biomass or hydropower. These both rely completely on water resources management.
- *Services*: For many service industries such as tourism—which is the fastest growing industrial sector of the world and among the key potentials in many developing countries—water, adequate sanitation and a healthy environment are elementary.
- *Health*: Water-related diseases kill far more than HIV/AIDS or conflicts together. This silent catastrophe accounts for a death—often of a child—every 10 seconds. This adds up to over 9,000 deaths a day and an incomparable loss of well-being and economy in the developing world where over 1 billion people lack appropriate water services and 2.6 billion do not have proper sanitation facilities, which means half the developing world do not have even a simple 'improved' latrine.
- *Economic growth*: It is necessary for poverty alleviation, but does not guarantee the same. Distribution of wealth is necessary. In economic terms, care must also be taken of not very profitable sectors such as (capital intensive) food production.

Monitoring the state and progress in the water sector in a country is not a simple task. It requires an interdisciplinary approach that may involve both qualitative and quantitative assessment techniques. These should be integrated to allow a wide range of issues to be addressed, while at the same time allowing the views and values of a range of stakeholders to be represented.

One approach to this demanding task was that by Lawrence et al. (2002), who ranked the world's nations according to the Water Poverty Index (WPI), which consisted of five components. Each component was based on several variables measuring different aspects of the water sector (Table 4). The maximum score from each component was 20 points, implying a theoretical maximum of 100 points for a 'water paradise' country.

Table 4 The structure of the Water Poverty Index (Lawrence et al. 2002)

Component	Data used
Resources	• internal freshwater flows • external inflows • population
Access	• % population with access to clean water • % population with access to sanitation • % population with access to irrigation adjusted by per capita water resources
Capacity	• PPP per capita income • under-five mortality rates • education enrolment rates • Gini coefficients of income distribution
Use	• domestic water use in litres per day • share of water use by industry and agriculture adjusted by the sector's GDP share
Environment	Indices of: • water quality • water stress (pollution) • environmental regulation and management • informational capacity • biodiversity based on threatened species

No paradises were found; the highest score, for Finland, was 78.0, and the lowest, for Haiti, was 35.1. Table 5 lists the world's macroregions (according to the classification of United Nations, 2006) with their overall rankings and scores from each component.

These index values are, of course, only very crude illustrations of the manysided and diverse role of water in the development of these regions. Table 1 and Fig. 11 are provided to give an overview of the situation in order to relate the other parts of the report to the magnitude and dimension of water sector challenges in different regions of the world.

It is interesting to compare the WPI with the basic demographic indicators:

- Population growth
- Growth of urban population
- Growth of aged population
- Share of aged population

In this analysis, the Human Development Index (HDI) is also included. In what follows, all of these data were standardised for zero mean and unit variance. This was done using the formula

$$z = (y - \underline{y})/s$$

Table 5 Regional values of the Water Poverty Index. Calculated from the country-specific results by Lawrence et al. (2002) and weighted by United Nations (2006) population data for the year 2005. The population column shows the region's share of the global human population, and the five following columns indicate the scores of the components of the Water Poverty Index as percentages of the world average (the higher the value, the lower the 'Water Poverty'). The WPI is the Water Poverty Index, HDI is the Human Development Index and 'Falkenmark' stands for thousands of m^3 of renewable water per capita per year

Region	Population	Resources	Access	Capacity	Use	Environment	WPI	HDI	Falkenmark
Europe	11%	109%	126%	129%	75%	120%	110%	0.8	10.4
North America	5%	131%	135%	126%	12%	154%	103%	0.8	16.7
Oceania	0%	162%	120%	121%	42%	115%	105%	0.8	57.9
East Asia	3%	92%	165%	138%	49%	115%	109%	1.1	2.6
China	21%	86%	83%	99%	118%	92%	98%	0.7	2.2
Southeast Asia	9%	129%	94%	103%	115%	98%	108%	0.7	12.3
South Asia	25%	87%	102%	90%	128%	93%	102%	0.6	2.6
Middle East and North Africa	5%	53%	124%	103%	104%	80%	96%	0.6	1.3
Sub-Saharan Africa	12%	98%	63%	70%	82%	91%	80%	0.5	7.1
Latin America and the Caribbean	9%	145%	105%	101%	88%	101%	104%	0.8	24.6
World	100%	100%	100%	100%	100%	100%	100%	0.7	7.6

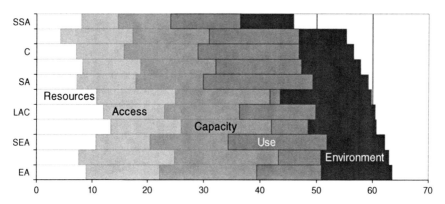

Fig. 11 Components of the Water Poverty Index by region, including the world average. Calculated after the data of Lawrence et al. (2002) and United Nations (2006)

where z is the standardised value (called z-score in statistics), y is the original value, $\bar{y}$ is the mean of this variable for all regions and s is its standard deviation. The results are presented as radar plots in Fig. 12.

We can see again that Europe's, North America's and East Asia's major demographic challenge is aging, and in these regions the HDI and WPI are relatively high, implying a good capacity to face challenges related to development and water. Latin America and the Caribbean, Southeast Asia and Oceania rank close to the global average in all respects considered in this analysis, and the population challenge is more manysided than in the high-income regions referred to above.

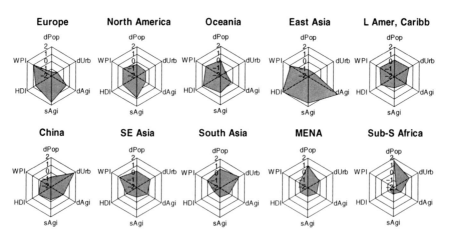

Fig. 12 Comparison of basic demographic indicators (dPop = population growth rate, dUrb = growth of urban population, dAge = growth of aged population and sAge = share of aged population) to Human Development Index (HDI) and to Water Poverty Index (WPI). Calculated after the data of Lawrence et al. (2002) and United Nations (2006). 0 means average of the regions, +1 means one standard deviation above the average, etc.

In China and South Asia, urbanisation dominates the demography, whereas in the MENA region and sub-Saharan Africa, population growth remains far above the world average. The latter is particularly low in capacity to deal with development and water-related challenges.

Impacts of Aging and Urbanisation on Water Resources

As water is a crosscutter of many facets of human life as well as economic, environmental and social systems, the impacts of demographic changes to water come through several avenues (Fig. 13). These avenues are to a large extent conditional to the economic system and phase of development in different regions of the world. In traditional rural economies, the impacts differ quite considerably from those in modern urban economies.

Water challenges will be related to aging principally through economic development and ability to be able to provide and maintain largely degraded water infrastructures, particularly in the field of water supply, sewerage systems and wastewater treatment as well as irrigation in some areas. The aging problem is a burning issue in rural areas where competent labour is scarce and insufficient.

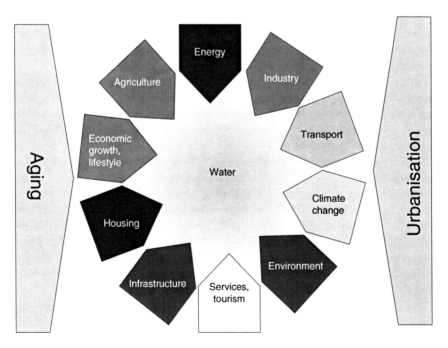

Fig. 13 Water is influenced by demographic pressures through several sectors and mechanisms

The Economic System: Industrial, Post-industrial, Traditional and Informal

When considering the impacts of global aging of the human population on the consumption of natural resources and particularly water, it is important to keep in mind some of the basic structural concepts of the global economy. This is because most of the impacts of aging propagate through the production and consumption systems of industrial and agricultural products and only to a smaller extent through direct consumption of water.

Industrial revolution in the eighteenth century meant a revolution in production systems and transportation infrastructure. Agricultural production was spatially disconnected from the consumption, and the same happened to industrial production. This also implied that water consumption and pollution were no more necessarily taking place close to densely populated areas. On the other hand, the energy and mass flows of nutrients, organic material and so forth were dramatically opened.

Extremely dense concentrations of human population were possible due to improved production and distribution systems, as were the parallel development of economic and financial systems, as well as governance systems that are democratic or based on technocratic authoritarian regime. International distribution of labour and large-scale industrial production units that are benefiting from economies of scale started to develop.

All these issues have, of course, been there in human societies for over a millennia, but it was no more that two centuries back that these systems started to undergo a revolution that allowed the modern, globalising economic system to kick off and the scales of operation started to soar.

As one of the consequences of industrial revolution, the towns and cities started to grow. This development has been progressing in Europe, North America and some other industrialising regions since the mid-nineteenth century. Urbanisation has since spread all over the world and is challenging the world's poorest regions in recent decades (Figs. 6–8).

The number of people that are directly dependent on the modern economic system has been in rapid growth ever since. However, the world economic system started to diversify, since by no means are all integrated in the modern economy.

The split of the economy into formal and informal sectors had already become evident in most African and Asian countries before the independence (Drakakis-Smith, 1987). A huge population with no adequate housing, health care and education lived very poorly, creating their own economic system that consisted of various illegal and semi-illegal activities. This informal, petty-commodity sector has grown to cover, in many cases, the majority of the population.

Today almost all governments are in favour of urban, industrial development. This urban bias boosts the urbanisation process. The hope has been that the growing industry and urban service sector would be able to absorb the excess labour force, but this has not happened, except to a small fraction of people. Even in the most successful countries in this respect, such as Brazil and Thailand, only one-fifth of the population are engaged in industry. The informal sector must absorb 20–70% of urban labour force (Todaro, 1997).

Even in countries undergoing rapid economic development such as China, the tax accumulation is far below the level of providing elderly care, pensions and health provisions. Therefore, the aged population is essentially dependent on their families or other informal mechanisms. With the expanding number of aged population and constant, if not shrinking, labour force, together with lessening of family ties within the modern-sector urban population, the risk of marginalisation of the aged population is growing rapidly.

The rural informal sector is usually not such a big problem socially as the urban one, although in many cases it is the major economic force. This is because many of the traditional, informal institutions function well, and social fabrics are there. The senior population maintains respect and authority and are not separated from the productive part of society as tends to be the case in the modern urban setting.

With the present rapid urbanisation, if social fabrics get partly or totally destroyed, the urban, informal sector runs into trouble. This is the case in particular when the urbanisation process is dominated by uncontrolled migration of excess rural people to urban areas.

Many strategically important cities have attracted foreign investments and industries in marked amounts since 1970. Such cities include São Paulo, Rio de Janeiro, Mexico City, Seoul, Taipei, Bangkok, Kuala Lumpur and many others. Infrastructure may develop rapidly, but not necessarily in a balanced way. Developments in the informal sector have been marginal, despite their rapid growth. Freshwater problems as well as other infrastructure shortcomings and environmental issues are very pronounced in the informal sector.

The second step of industrial revolution has been the outbreak of information technology in the past three decades. It has decoupled information flows from spatial settings and allows the distribution of labour in the service sector as well as in knowledge-based industries in an unprecedented way. This second phase of industrial revolution has profound impacts on natural resources use, too. It allows far better possibilities for delocalising production and service activities by outsourcing them wherever such activities are most economically provided. This development theoretically might loosen the urbanisation pressure but in practice most likely acts as an accelerator of urbanisation. But it might loosen the pressure of international migrations since the mobility of investments and other economic activities allows enhanced employment growth in developing countries, provided that they can create and develop an attractive environment for globalising economy. Countries such as the Asian Tigers, China and increasingly India have done this already and will develop their systems further. It remains to be seen whether regions such as MENA or sub-Saharan Africa will enter this scene or not.

The Club of Rome has been an important forum of thoughts and discussions on global sustainability issues over decades. One of the club's major outputs in the late 1990s was the F4 (factor four) concept (von Weizsäcker et al. 1997).

According to this concept, the technological and economic progress of industrialised countries has mainly been achieved by the increase in labour productivity.

A roughly 20-fold increase in 150 years has been possible through advancements in science and technology.

Resource productivity—GDP output per unit of resources such as water or energy—has not been constant, however. It first decreased in industrialised countries, but has later increased, but only at a rate of roughly 1% per year. In this perspective, China's recent progress is exceptional.

Resource productivity can be quadrupled using present knowledge. By halving the resource consumption, mankind is—in theory—capable of doubling well-being. In other words, resource productivity could be quadrupled. If resource productivity could grow by 3% per year instead of the present 1%, the F4 could be a reality in about 45 years, by the year 2050.

In addition, the emerging and industrialising economies, particularly in Asia, should leap over the polluting and resource-intensive phase of the industrialisation process, which has been characteristic of the development of North American, European, Chinese and other industrial economies. The emerging and developing economies would benefit themselves the most by building efficient economic structures.

Aging, Urbanisation and Household Size

Whereas economists and business people have been aware of the economics of scale for a long time and the world business environment has developed accordingly, this is not the case in private economies of people. In contrast, as people get wealthier, more urban and more aged, the household size goes down. In less developed countries, the average size of households was 5.1 persons in 1970, and it fell to 4.4 by 2000. In more developed nations it went down from 3.2 to 2.2 occupants (Keilman, 2003).

Smaller households—typical of urban and more aged societies—have a remarkably lower efficiency in resource use and thereby result in accelerated indirect pressure upon water resources. The water consumption per capita is not obviously too sensitive to household size, nor is the water that is consumed indirectly in food products. However, the increased per capita use of steel, concrete, wood and so forth adds to the strain on water resources due to the fact that water is inherent in the production of these materials. In the United States in 1993–1994, one-person households used 17% more energy per capita in comparison to two-person households (O'Neill and Chen, 2002).

Liu et al. (2003) analysed the resource-use impacts on decreasing household size in 141 countries between 1985 and 2000. The growth of household numbers has been considerably more rapid than the aggregate population growth. Even in cases in which population declined, the number of households grew. In aging and urbanising societies, the positive resource-use effect of decline in population growth may be offset by the higher per capita consumption in smaller households.

Macroeconomic Development and Demography

As the economically active share of the population decreases in relation to the aged population, the working population must support in one way or another the increasing masses of people who are in greatest need of various social and health services.

In more developed economies, the share of aged population will grow strongly, and employment will start to decline in 2010–2020. Tax accumulation for the funding of pensions and other services for the aged is already starting to be a problem, but the challenges will increase within 10 years. GDP growth will in most predictions increase at a slower rate, and the primary source of economic growth will be in the growth of productivity.

In countries in which the aging development is modest and urbanisation is not too massive (e.g. Latin America and the Caribbean, China, Southeast Asia) economic development may take place through both growing labour force and growing efficiency. Tax accumulation will be far easier to be kept high than in countries with a declining labour force. Relatively low need for aging-related public spending provides plenty of possibilities, but on the other hand, the countries in these regions typically have an immense need of infrastructure development.

In regions where fertility is still high and population growth remains rapid (Africa, West Asia, South Asia, Central Asia), it is critical to economic and social development just how the urbanising economy will absorb the huge incoming and typically relatively poorly educated labour force. Investment needs are enormous, and restructuring of the production system would be crucial to economic development, but both capacity and political vision are typically insufficient.

Globalising World Economy

The progress in information technology allows outsourcing of business activities internationally. The modern, urban sector is developing very rapidly in countries such as China, India and many other emerging economies. This includes not only industrial activities, but by and large also white-collar work by highly educated specialists. While in 2004 $10–15 billion of engineering services were offshored, the market is expected to grow to $150–225 billion by 2020. For instance, India's potential growth in providing engineering services might increase from the present US$1.5 billion up to even US$60 billion within that period (Booz Allen Hamilton, 2006). Countries that invest now on higher education, infrastructure and functionable business environments may attract considerable financial flow from these rapidly growing markets.

The concurrently often used split of the world's economies into the more developed and less-developed is clearly subject to profound changes within this process. The split within the countries among the globalised major urban centres and the rest of the economy may deepen, and, for instance, the local food markets may be disconnected from the soaring cash flows of the emerging urban middle class consumers, whose food supply may be increasingly provided by the international food market, and thus their consumer choices may influence the water resources in food-exporting countries to a far greater extent than today.

Globalisation seems to favour flexible and mobile components of the economy such as financial capital and knowledge. In contrast, the economic value of immobile and fixed assets such as uneducated labour and land goes down. This introduces an enormous challenge to the development of low-income countries, particularly their rural areas, in contrast to highly mobile and flexible urban-driven economies (see West Africa case below).

This demarcation respects less and less national borders. Rural areas tend to be drained of their most flexible capital—educated people—who migrate into cities in masses all over the world. The uneducated stay in the countryside. Not everybody, though. Many are pushed into cities because the rural economy does not offer a livelihood for all any more (see China case below).

The division of the urban population is between uneducated people merely 'pushed' from rural areas and those with an access to education and capabilities to benefit directly from globalisation strengthens. The latter class, if migrating from rural areas, tend to be 'pulled' to cities, rather than 'pushed'.

Primary, Secondary and Tertiary Cities

Whereas water issues of megacities of international appeal attract increasing attention (e.g. Lundqvist et al. 2005, Tortajada et al. 2006), the smaller urban centres may face more dramatic problems in their economic, social and environmental development. This naturally influences all aspects of the water sector in a big way.

Cities, above all the biggest ones, are the interface between a country or region and the globalising economy and culture. With their concentrations of capital, educated inhabitants and other resources, they may be linked more to each other than to their hinterlands. Owing to their economic links to the world market, they often prefer to purchase food and other commodities there if the market structure or prices cannot compete in the country or region in which they are located. This is obviously one outcome of the frequently made claim that nation-states are to a certain extent ceding influence in today's world (Hewett and Montgomery, 2001).

Somewhat smaller cities that are not the home of major export industries or government offices are also mushrooming these days.

In China, it is customary to classify important cities and towns into four classes according to their size and dimension of economic influence. Some characteristics of this classification are presented in Table 6.

The development gap between super cities and smaller urban centres may be several decades and tends to grow these days. This is due to financial shortage, and development of infrastructure has been by and large lower than planned and predicted by the government. However, there is a very high variation in performance and development, entailing that some lower category cities and towns are performing well, but many others have serious problems.

Table 6 Four categories of urban centres in China (Browder 2006)

Category	Class	Typical GDP per capita (US$)	Typical population size (millions)	Number	Total population (millions)	Wastewater treatment (%)	Water supply coverage (%)
1	Super cities	>3,000	>2	21	93	59	93
2	Medium cities	1,500–3,000	0.5–2	320	197	39	92
3	Developing cities	<1,500	<0.5	321	60	21	86
4	County capital towns	Less	Less	1636	96	11	82

Regional Features

This section discusses some crucial regional features related to the complex interplay of water, aging and urbanisation. This interplay takes quite different forms in different regions of the world and is conditioned by the level of economic development, the governance system, the level of integration to the world economy, cultural features and so forth, as the following examples from West Africa, China, India, Middle East and North Africa, as well as from Europe and United States, show.

Case 1: West Africa

Since the independence of West Africa in the late 1950s and early 1960s, the major pressures on the societies of the subcontinent have been 'exceptional population growth and brutal exposure to world markets', as articulated by Snrech (1998).

The societies were not prepared for this exposure in the colonial era. Accordingly, the societies have reacted in uncontrolled ways after independence. The local elites and governments have lost much of their power, and the informal sector booms. The most visible consequence has been the massive migration towards the coast, to urban areas; 3,000 towns and cities have absorbed 66 million new inhabitants.

The economies are based on agriculture and mineral commodity exports. Agriculture is gradually becoming increasingly market oriented, and the rural economy is depending increasingly on the market of food crops. At the same time, the integration of the agricultural sector to the world market has become stagnant, the priorities being in feeding the local population. Subsistence farming is still important. Many of the mineral resources have become very problematic in several ways, such as the diamonds of Sierra Leone, which feed corruption, warlords and international crime.

The manufacturing industry has not developed. While in South and Southeast Asia, the share of primary products of export earnings went down from 53% to 20% within the period 1970–1990, West Africa's development was quite different: the share decreased from 93% to a mere 89%.

The economies of West Africa are not competitive in attracting foreign investments to develop their industries, and the exports keep propagating very limited benefits to the societies as a whole. The region also loses market shares of traditional export goods, but is increasingly open for imported goods. The most severe issue is not the manufactured goods imported to the urban élites, but the massive dumping of subsidised agricultural products onto the world market, which destroys the development possibilities of local agro industries. The economies become increasingly indebted and are forced to concentrate on supplying the people with very basic needs.

West Africa is one of the most challenging regions of the world in its capability to face the water-related challenges within the coming few decades. The projected urbanisation expansion and looming aging boom will not ease these challenges.

In West Africa, the demographic matters—be it fertility, mortality, migrations, urbanisation or aging and taking care of the aged people—are strongly dominated by families, clans, kinship and other informal mechanisms, and this seems to be the case in most of sub-Saharan Africa. The demography and particularly the aging question are already impacted to certain level by the HIV/AIDS epidemic in West Africa, and most probably this disease will impact deeply the demographic structure of entire societies.

Case 2: China

China's industry is undergoing a rapid change from resource based to knowledge based. China is now witnessing the 'golden era' of economic development (or 'demographic window' of opportunity as called by Bloom and Williamson, 1998) as its urbanisation development is very rapid. The young labour force is relatively well trained, and the country has an attractive governance environment for investments. If China does not use this opportunity, it would have massive challenges in coping with the aging population within 10 or 20 years. China must be able to constitute a strong and competitive economy that can afford to generate enough tax revenues so that the country can cope with the rapid increase in the expenditure on the elderly population. This challenge is already materialising, but will start to grow massively within one decade.

China is a big country, and the provision of an environment for balanced development in the continent-size economy is somewhat a challenge (Varis and Vakkilainen, 2001).

Because most of the urban population growth is due to migration from rural areas, one efficient way of reducing urban problems—including those involving water—is to promote rural development. China is an interesting case in this respect. For several decades, the government restricted urbanisation, but recently the policy has been relaxing. At the same time two parallel processes are taking place.

On the one hand, the urban areas are witnessing unforeseeable economic growth that allows massive improvements in livelihoods and infrastructure; very large numbers of people have emerged from poverty. On the other hand, similar progress has not taken place in rural areas, and subsidies—which used to be very high—have been reduced. Rural areas are short of financial institution services, and other market infrastructure.

As a result, the expanding coastal megacities have started to import food from the world market, and the new urban wealth fails to trickle down to the rural economy. Urbanisation continues apace, and informal settlements have become considerable in size in Chinese cities. The sustainability of agro-ecosystems and rural livelihoods are also in question. Some recent studies reveal that the number of unregistered, mainly young, migrants in Chinese cities is increasing, which results in a rapidly growing informal sector (Söderlund et al. 2005; Varis, 2006).

Case 3: India

Whereas in the developed nations of Europe, East Asia and North America, care of the elderly will also in the coming decades predominantly be based on formal pension and insurance systems, in large parts of the world the informal, family and kinship-based systems will continue to dominate. Urbanisation changes this pattern rapidly in India (Dhar Chakraborti, 2004). Families have smaller number of children, the older people are more numerous and live longer, the multigeneration housing is becoming less and less a norm, and the societal status of the aged is eroding.

Older people are increasingly seen as economically unproductive, and the changing age structure will get in the way of capital formation. Family support to the elderly is declining rapidly in an urbanising society. India is one example among many in which a rapidly growing class of the poor is the aged, marginalised people.

Case 4: Middle East and North Africa

The MENA region is not only very scarce in water, but also subjected to a complex mix of development tendencies that make the future development of the water sectors of these regions particularly difficult. Varis and Abu-Zeid (2009) identified nine critical trends and tendencies that constitute a complex vicious circle:

- Population continues to grow and urban population doubles
- Rural water stress and poverty exist
- Economy is under structural pressures
- Regional integration is still low
- Education is under massive growth pressure
- Most new jobs are informal
- Food security is increasingly based on self-reliance instead of self-sufficiency
- Climate change may cause prolonged droughts
- Environmental stress should be relieved

These tendencies are profoundly interlinked with one another. In a way, they constitute a set of vicious circles that partly accelerate one another. The water sector has obvious entry points to all of the main components of the system.

For a growing share of the MENA population, the informal sector becomes increasingly familiar, particularly in the urban areas. Most of the heavily recommended issues such as decentralisation, cost recovery, economic instruments, legislation, private sector involvement, stakeholder participation, adaptation of costly, non-conventional water technologies such as recycling and desalinisation, and many others have often totally different shades in the informal sector in comparison to how they appear in the formal sector.

Another source of potential surprises and challenges to the water sector of the MENA region comes from political factors, both within the countries and in the international arena. The past decades have seen long stable periods in several of the region's countries, but also conflicts and unpredictable changes in regimes and political-economic systems.

The water sector documents have so far been too immune to the consideration of such factors and their potential roles in shaping future water strategies for the region. The following list includes some potentially very important future topics of water policies, which are underrepresented in much of the water sector debate today (see for instance CEDARE, 2004, 2005; CONAGUA, 2006):

- From rural to urban. How could the water sector target the doubling urban population? The number of urban poor may still essentially grow faster. Will slum-upgrading policies become a crucial task for the MENA region's water sector? How will the water sector developments influence the uncontrolled rural exodus, and are there even other important connections to demography?
- From natural resources to human resources reliance. How could the extremely scarce freshwater of the region be used to create more economic and social welfare than today? The modern industries and trade are still in their infancy in the MENA countries. Scarce natural resources are becoming increasingly under pressure, but the huge human potential should be tapped to develop the region. Basing the economy on human resources and modern industries instead of continuing on the exceptionally high reliance on natural resources, in which the region has a growing competitive disadvantage, should be an issue of the water sector.
- From physical scarcity to environmental quality. Water quality and pollution abatement definitely deserve increasing concern. Water quality and environmental protection of water resources have largely been given lower priority in terms of investments, without considering the associated environmental and health degradation costs. The MENA region cannot afford for this to continue.
- From national to regional policies. The MENA region will most likely be exposed to increasing trade and globalisation pressures. The water sector will be linked progressively more to economic policies, and the regional dimension is likely to grow. The future of transboundary water issues that are already very high on the region's political agenda is still unclear to predict, but may have the potential of resolution if more regionally focused water policies see daylight.

Accordingly, the pressures and potential surprises to the water sector come largely from trends, tendencies and occurrences that are external to the water sector. Demography is one of the most pronounced of these, since the MENA region is subjected to rapid population growth, extreme urbanisation pressure and emerging aging problems. In the coming few decades, the water sector must be prepared to tackle the exceptional development pressures which will face the MENA region.

Case 5: Europe and the United States

Within the European Union (including the EU member states as of 2006), the total population is expected to fall slightly by 2050, but the working population will then be 17% smaller than now. The elderly population (65+) will grow by 77% during the same period. Economic growth rate is projected to halve from the present level, and the public spending requirements for the pensions and health care for the senior citizens poses an immense strain to the economies of the EU member states (European Union, 2006).

As an increasing share of financial flows goes to social care, the service sector flourishes but capital formation is challenged in the national economies, and savings as well as investments on productive activities will decrease. The international economic competitiveness will be seriously challenged, and capital from high-income regions with very aged population is at risk of escaping increasingly into countries with cheaper labour costs (including less taxation pressure). Labour participation by older households is increasing rapidly in Europe and the United States. This obviously offsets some of the aging-related issues, as the working life-span increases.

If compared with Europe, the aging challenge of the United States is not nearly as great. The United States is expected to double its population size by the end of this century. The country has one of the most favourable age structure among industrialised countries, and the labour force is expected to grow at least until 2050 (McDonald and Kippen, 2001; Dalton et al. 2008). Even though the aging of the US population is not as dramatic as in Europe or East Asia, the factors related to tax formation, social security and so forth will resemble much of those in Europe, but will come later and not be as serious as in the 'old continent'.

The United States has some interesting demographic patterns that have caught the attention of the American Water Works Association as being among the key trends affecting water utilities in the coming decades (Means et al. 2005, 2006). First, most population growth occurs in arid or semi-arid areas of west and south, most rapidly in water-scarce states such as Arizona, Nevada and Colorado. The immigrating, relatively young population is the major force to the population growth. Second, the slowest population growth occurs in some water-rich areas in northeast and midwest states.

The latter ones will probably face very similar aging-related challenges to those in Europe (low tax accumulation, aging infrastructure), and the former ones will face physical water-scarcity problems, with primary reason being expansion of population and economic activities, and the aging will be of less importance.

A Note on Uncertainty of Population Projections

No population prediction for several decades ahead is free of major uncertainties. So is the case with the United Nations (2006) population prospects used in this analysis. The forecast includes a set of scenarios that shed light on the uncertainty range with which these prospects operate.

When it comes to the world's total population, the uncertainty range for the 2050 prediction is around ±17%. This range applies with small variation also to predictions for the continents.

Most of this variation is due to uncertainties in fertility. The birth rate has been proven in the past to be difficult to forecast and therefore most of the uncertainty in the decade-long population predictions can be attributed to fertility predictions. The United Nations Population Bureau actually bases its different population scenarios (low, medium, high, etc.) on different assumptions for fertility rate. In those scenarios, mortality rates are not varied across different scenarios except the HIV/AIDS scenario which anyway is highly hypothetical and does not provide much aid in assessing uncertainties in the mortality rates.

When it comes to the uncertainties of the population size of the aged population, in a timeframe of four to five decades ahead, such predictions are much more accurate than the predictions of the size of the youth population or even the total population. This is simply due to the fact that all the individuals that will reach the age of 65 before the middle of this century have already been born. The uncertainty that remains is due to either mortality estimates or mobility of people. The former is far more significant in continent-wide aggregate data such as used in this study.

The uncertainty in the mortality rate differs significantly across continents, being clearly highest in Africa and lowest in high-income regions. Whereas the life expectancy has been growing rapidly in all other regions, it has declined in sub-Saharan Africa. This is caused by the spread of diseases such as HIV/AIDS, malaria, tuberculosis and others, as well as the growing incidence of armed conflicts.

Whereas the United Nations Population Bureau does not give uncertainty estimates for the mortality assumptions it is using in its scenarios, neither does it include such uncertainties in its population projections, and it is not feasible in this study to provide such information. However, it can be derived from what was discussed before that the aging projections are relatively accurate, even decades ahead in areas where the morbidity of the population is not expected to undergo major surprises. Projections for Africa might be far more uncertain than for the other continents due to the HIV/AIDS pandemia. The projections for the last decades until 2050 are naturally subjected to growing uncertainties, but as has been said, the population that is going to reach 65 years of age by mid-century is already born, and from past experience, the mortality rates can be predicted essentially more accurately than fertility rates, and thereby the aging predictions several decades ahead entail relatively little uncertainty.

Epilogue

The world's population is undergoing two major shifts during the first half of the twenty-first century. These are urbanisation and aging. Both of these transitions take place in all parts of the world, and in most of them they both reach dramatic dimensions. Water, aging and urbanisation constitute a complex entity. These demographic factors influence water indirectly in a number of ways, which are often interrelated. Figure 13 indicates the key avenues through which aging and urbanisation are related to water. This chapter scrutinised these aspects in different geographical regions of the world. The world is very rapidly becoming more urban and more aged, and the water sector should appreciate this profound change.

Acknowledgements The author acknowledges the excellent support provided by the Helsinki University of Technology, Water Resources Laboratory and its staff. Thanks are particularly due to Professor Pertti Vakkilainen for the encouragement on this work. Inspiration and comments from Professor Asit K. Biswas, Dr. Cecilia Tortajada, Professor Wolfgang Lutz and Professor Malin Falkenmark are greatly acknowledged. The financial support from Maa- ja vesitekniikan tuki r.y is greatly appreciated.

References

Bloom DE, Williamson J (1998) Demographic transitions and economic miracles in emerging Asia. World Bank Econ Rev 12(3):419–455
Booz Allen Hamilton (2006) Globalization of Engineering Services. NASSCOM/Booz Allen Hamilton, New York
Browder G (2006) China Urban Water Sector Study. World Bank, East Asia Urban Development Unit, Washington D.C
CEDARE (2004) State of the Water in the Arab Region. Arab Water Council and Centre for Environment and Development for Arab Region and Europe, Cairo
CEDARE (2005) Status of Integrated Water Resources Management: IWRM Plans in the Arab Region. UNDP, Arab Water Council and Centre for Environment and Development for Arab Region and Europe, Cairo
CONAGUA (2006) Middle East and North Africa: Regional Document (4th World Water Forum). Comisión Nacional del Agua, Mexico City
Dalton MG, O'Neill BC, Prskawetz A, Jiang L, Pitkin J (2008) Population aging and future carbon emissions in the United States. Energ Econ 30:642–645
Dhar Chakraborti R (2004) The greying of India – population ageing in the context of Asia. Sage Publications India, New Delhi
Drakakis-Smith D (1987) The third world city. Routledge, London
Ergüden S (2005) As developing cities grow, where do the elderly people go? In: AARP Global Report on Aging 2005: 14–15
European Union (2006) The Impact of Ageing on Public Expenditure: Projections for the EU25 Member States on Pensions, Health Care, Long-Term Care, Education and Unemployment. European Policy Committee and the European Commission's Economic and financial Affairs Directorate General, Brussels
Hewett P, Montgomery M (2001) Poverty and public services in developing-country cities. Policy Research Division Working Paper no. 154 (New York: Population Council)
Keilman N (2003) The threat of small households. Nature 421:489–490
Lawrence P, Meigh J, Sullivan C (2002) The Water Poverty Index: An International Comparison. Keele Economic Research Papers 2002/19, 24 p
Liu J, Daily GC, Ehrlich PR, Luck GW (2003) Effects of household dynamics on resource consumption and biodiversity. Nature 421:530–533
Lundqvist J, Tortajada C, Varis O, Biswas A (2005) Water management in megacities. Ambio 34:269–270
McDonald PM, Kippen R (2001) Labor supply prospects in 16 developed countries, 2000–2050. Popul Dev Rev 27(1):1–32
Means EG, Ospina L, Patrick R (2005) Ten primary trends and their implications for water utilities. J Am Water Works Assoc 97(7):64–77
Means EG, Ospina L, West N, Patrick R (2006) A Strategic Assessment of the Future of Water Utilities. American Water Works Association, Denver, Colorado
O'Neill BC, Chen BS (2002) Demographic determinants of household energy use in the United States. Popul Dev Rev (Suppl.) 28:53–88

Snrech S (1998) West Africa undergoing long-term change. In: Cour J-M, Snrech S (eds) Preparing for the future–a vision of West Africa in the Year 2020. Club du Sahel, OECD, Paris

Söderlund L, Sippola J, Kamijo-Söderlund M (eds) (2005) Sustainable agroecosystem management and development of rural-urban interaction in regions and cities of China. Agrifood Research Reports 68 (Jokioinen, MTT Agrifood Research Finland)

Todaro M (1997) Economic development, 6th edn. Longman, London and New York

Tortajada C, Varis O, Lundqvist J, Biswas AK (eds) (2006) Water management for large cities, 18–44. Routledge, London

United Nations (2006) World Population Prospects: The 2004 Revision. United Nations Population Bureau, New York

Varis O, Vakkilainen P (2001) China's 8 challenges to water resources management in the first quarter of the 21st Century. Geomorphology 41(2–3):93–104

Varis O (2006) Megacities, development and water. Int J Water Resour Dev 22:199–225

Varis O, Abu-Zeid K (2009) Water management to 2030: Trends, challenges and prospects for the MENA Region. Int J Water Resour Dev (forthcoming)

Varis O, Biswas A, Tortajada C, Lundqvist J (2006) Megacities and water management. Int J Water Resour Dev 22:377–394

von Weizsäcker EU, Lovins A, Lovins H (1997) Factor four: Doubling wealth, halving resource use. Earthscan, London

World Bank (2003) World Development Indicators. The World Bank, Washington D.C

Water and the Next Generation – Towards a More Consistent Approach

Malin Falkenmark

Introduction

The water cycle is the bloodstream of both biosphere and human society. Every human body contains some 70% water, which has to be partly renewed every day. The body does not function when the water content diminishes too much. Water – although chemically simple – is a highly complex substance with many different functions: health, income generation, energy production, biomass production, habitat and carrier functions.

Problematique

This complexity makes water management intricate. Strong driving forces now influence human–environment relations. The technical–economic perspective that has been dominating in the past is no longer enough. It is being increasingly clear that many tasks remain undone: huge problems remain, primarily ecosystem degradation and water pollution.

In the future we may expect water to be deeply involved in the ecological overshoot problematique (WWF, 2006). Although in different roles, water is involved in all of the four categories of factors identified by Diamond (2006) as linked to societal collapses of civilisations:

- Destruction of natural resources: natural habitats, biodiversity, soil degradation. Water functions involved: habitat of aquatic ecosystems, erosion, salinisation.
- Natural resource ceilings: freshwater availability, photosynthetic capacity. Water functions involved: water-cycle-based freshwater renewal, biomass production (food, bioenergy).

M. Falkenmark (✉)
Stockholm International Water Institute (SIWI),
Drottninggatan 33, SE-111 51, Stockholm, Sweden
e-mail: malin.falkenmark@siwi.org

- Harmful products. Water functions involved: water as carrier towards food chains.
- Population issues: population growth, water as a human right. Water functions involved: increasing water requirements, both blue and green water.

New tasks are emerging: how to feed the world under escalating water scarcity. Also, the efforts to turn world economy away from fossil fuels towards renewable energy sources, in particular bioenergy production, will demand huge amounts of water. Two great challenges are adding to the complexity: on the one hand the fact that population growth is concentrated in tropical and subtropical regions and on the other the ongoing climate change which will be adding to the resulting thumbscrew effect on the world's freshwater resources.

Great regional differences characterise the problem constellations in different regions: in temperate climate countries particularly quality deterioration and dying water bodies like the Baltic Sea and the Mexican Gulf; in arid developing countries water supply and sanitation deficiencies, droughts, poor crop yields, and upstream–downstream conflicts, of particular relevance to transnational rivers.

With escalation of water challenges, past simple management approaches are getting increasingly insufficient. What is particularly urgent is to incorporate linkages in our understanding: both quantity/quality linkages, quantity/ecosystem linkages and quality/ecosystem linkages. For instance, irrigation has resulted in large-scale river depletion in many river basins, now covering 15% of the continental land (Smakhtin et al. 2004). This means that overappropriation of water is a new type of problem that has become evident in the 1990s. The resulting river depletion has brought rainwater into new focus, generating growing interest in rainwater harvesting and rainfed agriculture.

Broadening the Perspective

A shift in thinking has turned out to be necessary (Falkenmark and Rockström, 2006) by drawing new attention to the water 'hidden' in the soil, the so-called green water resource. Although this water was thoroughly addressed in the 1970s by the famous Soviet scientist, Professor M.I. L'vovich, global interest did not take off until the concept 'green water' was introduced at an FAO seminar in January 1993 in Rome (FAO, 1995) (Fig. 1).

Attention is also required to water's destiny after use: whether consumptive (evaporating) water use or water use that is throughflow based and results in a return flow carrying a content of pollutants. Water use for domestic and industrial purposes is basically a throughflow-based use along an intake/use/outflow chain, often picking up pollutants during use, leading to river pollution, and calling for wastewater treatment to reduce contamination.

What brought the consumptive dimension of water use into focus has been river depletion appearing in the wake of increasing irrigation: does water evaporate during use or not? Such depletion will reduce the return flow that carries pollutants back to the river system and the water available for use downstream. The fact that irrigation

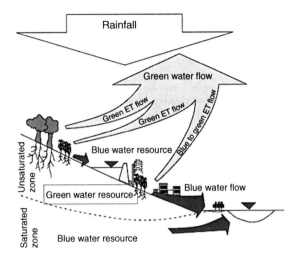

Fig. 1 While precipitation can be seen as the true water resource, it is over land partitioned between the naturally infiltrated water in the soil (green water) accessible for plant production and the liquid (blue water) passing through rivers and aquifers and accessible for direct societal use

is often a wasteful use of water in the sense of very low water efficiency has led to a call for 'more crops per drop' of either water withdrawn or water evaporated. However, to the degree that 'crop-per-drop' efforts will reduce the blue water losses in terms of channel bottom percolation and field drainage, river depletion will increase even further – a physical fact that has not been well understood.

Hydropower production is a mainly non-consumptive water use, except for the evaporation from water storages involved to facilitate production during the time most needed. Flow control demands reservoirs, the size of which depends both on the streamflow variability and on the size of the river. Many conflicts of interest and much opposition to flow-control infrastructure are due to effects on aquatic ecosystems and resettlement needs.

Water for biomass production is an area now attracting growing interest in view of the large degree of undernourishment and the rapid population growth that tends to be concentrated to semi-arid areas. Taking a global generalised perspective, most of the consumptive water use is in fact linked to the terrestrial ecosystems, consuming altogether some two-thirds of the rain over the continents. The new interest paid to bioenergy adds to the relevance of this problematique, since bioenergy production may generate larger income to farmers. Biomass production involves green water and is an issue of water uptake in the root zone and relation transpiration/vapour flow, carbon dioxide being the other.

All this boils down to three sets of problems:

- Quantity limitations, generating competition for water
- Quality degradation, reducing water usability and increasing ecosystem deterioration
- Problems originating from side effects of landscape manipulations linked to socio-economic development, generating ecosystem impacts

Based on these different challenges and considerations, the rest of this chapter will address the water-related challenges of the next generation.

Resource-Constraint Realities

Principally, the basic water resource in an area is precipitation plus inflow. That water can be used in two contrasting ways:

- As soil moisture/green water for – besides natural vegetation – plant production for food and bioenergy. This water use is a consumptive or depletive use in the sense that the water is vaporised during use, as a consequence reducing runoff generation.
- As liquid water/blue water after water withdrawal for water supply of households, municipalities and industry and for irrigation. The former types are mainly throughflow-based uses, returning the water after use as wastewater. The latter type is depletive except for any surplus that returns to the system as return flow carrying agricultural chemicals.

Depletive Water Requirements

Food: Photosynthesis starts with the splitting of the water molecule. Biomass production involves green water and is an issue of water uptake in the root zone and consumptive water use via transpiration. With current water productivity in agricultural production, on average some 0.5 m^3 of water is consumed/vaporised in producing the equivalent of 1,000 kcal vegetal food (Falkenmark and Rockström, 2004). Producing animal-based food is even more water consumptive due to low energy efficiency in meat production and some 4 m^3 of water is consumed per 1,000 kcal animal-based food.

The amount of food required per person varies with size, age and degree of physical activity. Although the food requirement per capita is lower, FAO assumes that by 2025 the apparent food-consumption level – that is, what is to be produced to secure minimisation of undernutrition – in developing countries will have risen to close to 3,000 kcal/person/day. The amount of water that will be consumed in producing that amount of food would amount to 1,300 m^3/person/day. This is 70 times more than the so-called basic need, seen as necessary for drinking and household purposes in developing countries.

Assessing the amount of water that will be required if this amount of food should be produced for every person in the developing world indicates that, if hunger is to be alleviated by 2030, an additional 4,160 km^3/year would be required, and for 2050 growing to 5,160 km^3/year (Rockström et al., 2007). This is almost twice as much as was consumed in the mid-1990s (Fig. 2). The next question is how these huge water requirements could be met. Recent assessments of how much irrigation could be expanded indicate a rather limited potential, with best guesses being in the interval 300–600 km^3/year (IWMI 2006). The rest will have to come from improved rainfed agriculture. The potential may be large in view of the considerable losses typical for

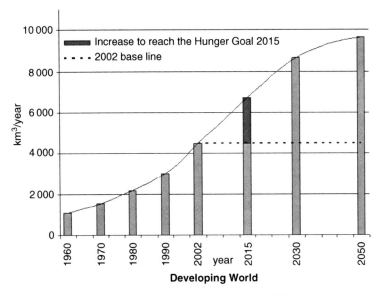

Fig. 2 Overall water requirements for food production in 92 developing countries to alleviate hunger. From SEI (2005)

semi-arid smallholder farming in sub-Saharan Africa. Rainfall, if better used, might in fact allow considerable yield (Rockström et al., 2007).

From a purely water perspective, it would be possible, even in semi-arid savanna regions, to attain crop yields many times higher than the yields experienced at present in smallholder farming, generally around 1 ton/ha only (Rockström and Falkenmark, 2000). Figure 3 clarifies the different water losses, both blue in terms of surface runoff and percolation to groundwater and green in terms of evaporation from wet soil, that together explain the large gap between actual and potential crop yield.

Water productivity can in other words be much improved, particularly in the semi-arid and countries in the savanna zone where undernutrition dominates. This means that large evaporation losses in smallholder farming can be avoided by protecting the crops from root damage during frequent dry spells by protective irrigation, based on rainwater harvesting. This would allow increased yields but reduce the additional amount of water required by 2030 from 4,160 to 3,000 km^3/year. In the following 20 years, continued productivity improvements might compensate the water requirements for the additional population.

Bioenergy requirements: In addition to the need for increased crop production to feed a growing humanity, the energy sector will also require huge additional amounts of water for the production of bioenergy, supposed to reduce the energy crisis and by getting away from fossil fuels (Berndes, 2006). An expansion of energy crop production to scales indicated by a study by IIASA/WEC might introduce a new appropriation of ET that can be as large as present global crop production.

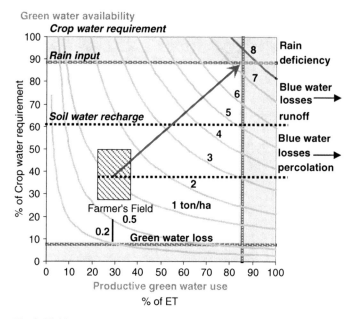

Fig. 3 Yield gap and soil-water availability in typical smallholder farming in sub-Saharan Africa, indicating large green and blue water losses. Modified from Rockström and Falkenmark (2002)

Impact of additional consumptive use, to be reflected in further decreasing streamflow in a basin, might originate both from withholding of rainwater from runoff generation by increased ET and from plantation irrigation turning blue water green. Although the incidence of irrigation is difficult to project according to Berndes, it might lead to a substantial additional withdrawal/consumptive use during this century. If 15% of the production was provided through irrigation with 50% efficiency, this would add up some 1,200 km^3/year by 2050, equivalent to more than half the current level of agricultural irrigation withdrawals.

Blue Water Constraints

There is a popular saying in the water community that it is not water that is scarce but good management approaches. This is unfortunately not altogether true: water is getting seriously scarce in arid regions of the world where population continues to grow. IWMI (2006) has used the concept 'physical water scarcity' as opposed to 'economical water scarcity'. The former refers to use to availability beyond 75%, the latter below 25%. Recently, Rijsberman (2006), focusing on densely populated arid areas in Central and West Asia and North Africa, stressed that 'water will be a major constraint for agriculture in the coming decades, and particularly in Asia and Africa this will require major institutional adjustments'.

The fact that blue water availability is limited and that depletive blue water use reduces the resource suggests the existence of a certain carrying capacity (c.f. Diamond's natural resource ceilings, 2006). The author of this chapter had already raised the issue of an approaching water scarcity crisis in the mid-1980s (Falkenmark, 1984, 1986). In the mid-1980s, Falkenmark empirically suggested that countries with more than 600 p/flow unit (p/f u) of 1 million m^3/year showed signs of water stress and those beyond 1,000 signs of chronic water scarcity. The level 2,000 p/f u was suggested as the water barrier for a water-dependent advanced country since repeated reuse tends to generate groundwater salinity.

Now, some 20 years later, the strong driving forces in terms of population growth envisioned socio-economic development give a special reason to analyse the situation again. What can be foreseen for the period up to 2050 when the population is supposed to have stabilised?

The terminology around water scarcity and water stress is unfortunately confusing and needs to be more precise, distinguishing between *water scarcity*, referring to the relation between availability and demand, *water stress*, referring to management problems, and *water shortage*, referring to water deficiency.

The term *water stress* is being widely used for example by UN. It is basically a technical water stress that refers to mobilisation ratio, conventionally expressed as use-to-availability ratio, also referred to as criticality ratio (Alcamo et al. 2000). As this ratio increases, more and more infrastructure is required. In Europe, according to Balcerski, the costs had increased greatly even as the ratio merely passed 20%. In the UN system, more than 40% is considered high water stress. Beyond 70% can be seen as overappropriation, since a residual streamflow is required for protection of aquatic ecosystems.

The population-driven water scarcity should be referred to as *chronic water shortage* for situations of water crowding, linked to how many people have to share the available freshwater resource. This is referred to by an indicator, expressing number of people sharing each flow unit of blue water, or inverted as per capita water availability. The empirically based severity intervals introduced by Falkenmark, discussed above, often referred to as standard indicator, are still being broadly used in the literature: water shortage, originally referred to as 'water stress', beyond 600 p/f u of 1 million m^3/year (inverted 1,700 m^3/person/year), and chronic water shortage beyond 1,000 p/f u (inverted 1,000 m^3/person/year) (Falkenmark, 1989).

Figure 4 illustrates this two-dimensional approach to water scarcity: water stress/use-to-availability ratio on the vertical axis and water shortage/water crowding on the horizontal axis (Falkenmark, 1997). Combining the 40% use-to-availability ratio limit with the 1,000 p/f u water-crowding limits gives four main realms for water scarcity.

Diagonal lines can be entered for different blue water demand levels (Falkenmark and Lannerstad, 2005). Two driving forces are at work in the diagram: population growth pushing a basin to the right and increasing water demands pushing it vertically. Where in the diagram a particular country is located gives an indication of the degrees of freedom in terms of how large water demand the country can afford. When population is increasing, a country will be moving up along the diagonal line

Use-to-availability ratio	Less than 1,000 p/f u	More than 1,000 p/f u
More than 40%	High water stress	Severe water shortage
Less than 40%	Moderate or low water stress	Chronic water shortage

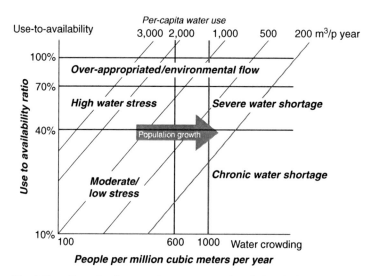

Fig. 4 Two-dimensional approach to water scarcity, distinguishing between water stress, linked to use-to-availability ratio, and physical water shortage or water crowding, linked to population pressure on water availability. *Diagonal lines* indicate per capita water use. Since the water beyond 70% is needed for aquatic ecosystems (so-called environmental flow), withdrawals beyond that level involve overappropriation. From SIWI (2007)

if per capita demands are to remain unchanged and in the horizontal direction if no further infrastructure development will be acceptable. The 70% line indicates the constraints in terms of maximum acceptable water withdrawals, provided 30% is to remain as residual streamflow (so-called environmental flow, Smakhtin et al. 2004) for protection of aquatic ecosystems.

Assuming that the minimum basic withdrawal for municipal water supply and industry would be 200 m³/person/year, the 200 m³/person/year line shows what might be required for domestic and industrial use only. Beyond that line there is, in such a case, no surplus for irrigation – only rainfed agriculture is possible. The most critical predicament appears where the diagonal line hits the 70% level. Further population growth will imply difficulties even to find enough water for population and industry, and desalination or alternatively water import from some neighbouring basins probably develops as the only realistic solutions.

Analysis will show that when water crowding has passed 3,500 p/f u, irrigation will not be possible if a 30% environmental flow reserve is to be conserved. When water crowding is beyond 5,000 p/f u, all blue water will be needed for domestic and industrial use only.

Trying to identify where those vulnerable countries are today in the diagram, one has to count backwards by incorporating the expected population growth till 2050, which is somewhere between 35% and 95% for most developing countries. The conclusion is that the regions with the most severe crisis in terms of water shortage are those now on some 2,000 p/f u and with rapid population growth. Figure 5 shows the situation for a number of river basins in the United States, China and India. Basins approaching a severe water shortage crisis are Limpopo, Yellow, Hai, Huai and Cauvery rivers.

Depending on where in the diagram a certain country/region or basin is currently situated, this approach will also allow a judging of what sort of action has to be taken to secure adequate food production and adequate infrastructures:

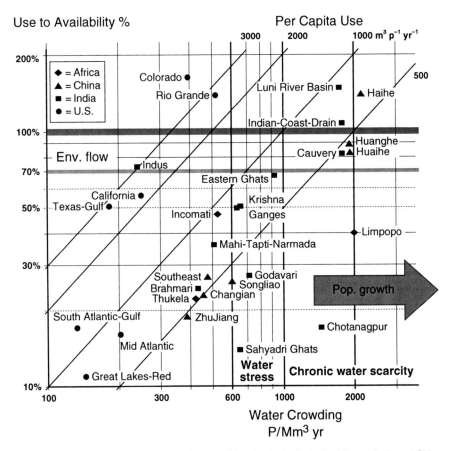

Fig. 5 Two-dimensional water scarcity in a set of river basins in the United States, India and China. Modified from Falkenmark and Lannerstad (2005)

- Storages to allow flow control to secure water in dry season
- Irrigation expansion
- Upgrading of rainfed agriculture
- Regulations regarding residual streamflow (environmental flow reserve)

Pollution-Avoidance Imperative

Hydrocide Threat

After three decades of water quality management efforts, the inability all over the world to halt water pollution remains a serious failure. The crisis of water pollution is increasing steadily in the developing world, making hydrocide an approaching reality (Lundqvist, 1998). Downstream stakeholders are increasingly being left without usable water. In industrialised countries, the wilful neglect (Lundqvist, 1998) of problems caused by wastewater disposal practised by many industrial branches operating in temperate climate was dampened by access to plenty of dilution water available in the rivers. The transfer of those models to developing countries in tropical regions with a long dry season has however been a major mistake (Falkenmark, 2005). Action is now urgently required before water quality degradation makes it impossible for such countries to get out of the poverty trap because of lack of usable water.

The scale of water pollution in the world is huge. In a thought-provoking model study, Simonovic (2002) has demonstrated how the development of persistent pollution (PCB, DDT, dioxin and hundreds of others, many of them hormone disruptors), as assumed by the water blind Club of Rome, might spread through water-related feedbacks and influence human health through drinking water and food. We know from other sources that persistent pollution can be magnified millions of times in the food chain, that humans are feeding at the top of the food chain, that the pollutants accumulate over time in body fat of living creatures, that male sperm counts have diminished dramatically since the 1940s and that women transfer pollutants stored in their bodies for many decades to their foetus/child during gestation and breast feeding (c.f. the state-of-the-art publication *Our stolen future* by Colborn et al. 1996).

Simonivic's model indicated that this pollution may come to have dramatic impacts on world population. In his most extreme scenario, he relied only on river dilution (assumed no extra wastewater treatment) and was able to show that the result might in fact be a population shock around 2040, dramatically reducing the world population, which after stabilisation would be around 3 billion (Fig. 6, curve b). In other scenarios, he analysed the effects of different degrees of water pollution abatement efforts. Assuming a doubling of current wastewater treatment and addition of some extra dilution water generated through desalination, he was able to reduce the population shock around 2040 so that the population would stabilise after recovery at around 6 billion.

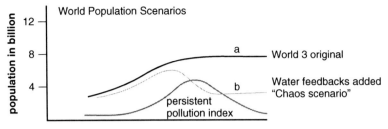

Fig. 6 Water-mediated impact of persistent pollution on population, comparing Club of Rome's World 3 Model original (*curve a*) and the same model, adding Simonovic's water-related feedback (*curve b*). Data from Simonovic (2002)

What these scenarios suggest is that persistent pollution is a global-scale issue of fundamental importance that has to be urgently dealt with in order to avoid a threatening hydrocide a few decades from now. Solutions include clean production, rapid phasing out of persistent chemicals, water reuse, wastewater treatment and, where realistic, desalination to compensate severe water shortage.

Even if extreme, Simonovic's study may be taken as a warning signal that pollution abatement is urgent. Policies have to be put in place in terms of pollution source abatement for both domestic sources: industrial and agricultural sources of pollutants.

Domestic Sources Abatement

The future avoidance of domestic pollution should have three parallel goals: to minimise disease agents, to facilitate wastewater reclamation for recirculation and to facilitate nutrient reuse. We know that poorly planned latrines – where insufficient attention is paid to hydrologic conditions – may generate groundwater pollution and make urban areas unsafe and an insecure place for residence and human activity. The situation indicates the need for modified toilets (ecological sanitation) and, especially in water-scarce regions, dry sanitation. Such sanitation refers to urine-separated toilets, with no or little water added, and is an elegant way to minimise water use and to recover usable fertilisers from both urine and solid faecal matter.

Wastewater may be seen as a resource in terms of not only water but also nutrients that might be marketed after reclamation. City wastewater can be reused both in the city and in the peri-urban area as a potential resource for food production benefiting both farmers and the urban market. While nutrients can be safely recycled from the wastewater, ecological sanitation might be beneficial, not only for poor developing countries but also for developed countries. Human faeces can be seen as a resource that can be safely reused after reduction of pathogens and heavy metals. The potential risks involved may not yet be well understood, however.

Industrial Production Sources Abatement

Minimising industrial pollution sources will also involve several parallel goals: both to avoid output of heavy metals and other hazardous substances and to facilitate wastewater reclamation for recirculation. But there may also be deposits, tailings and waste dams from past activities that have to be rendered harmless.

In poor countries, small-scale industry poses severe problems: while they are fundamental to raising income, they are too small to manage pollution control techniques developed in the West. Especially, water pollution abatement initiatives in small- and medium-scale industry may involve considerable complexity (Agarwal, 2002). A typical remediation situation in India may involve several hundreds of tanneries along a small river, for which a common effluent treatment plant is being proposed. Even if public awareness is high and expectations are great, financial and institutional dilemmas constitute huge obstacles. The industries may in fact represent the main source of employment in the area. A typical situation in China with the aim to achieve a substantial reduction of industrial wastewater, may even involve the closing down of more than 10,000 industrial units. Problems also remain in terms of low environmental awareness and a polluting economic structure of industry.

Agrosource Abatement

In agriculture, which represents the main source of nutrient pollution, the excessive loading has caused worldwide eutrophication and the formation of oxygen-free coastal zones around the world's oceans. Experience during past decades shows that agricultural pollution and especially eutrophication are very difficult to control on a larger scale. The reason is that agriculture is such a fundamental activity in a country, and fertilisers are central for good yields. Pollutant sources are spread all over the agricultural landscape, which means that combating this source has to be performed by almost every single farmer himself. The abatement goal must also include reduction of pesticide output by turning to biodegradable herbicides; improved land-use practices and fertilisers; and enhancement of local nutrient recycling by feed production in the same regions where cattle is raised and the meat customers are located.

Linkages to Ecosystem Protection

The third major challenge has its focus on ecosystem protection. Ecosystems are living components in the physical environment, building up the life support system for all living creatures. This makes their destruction through human activities hazardous for humanity, not only for those living today but also the coming generations. However, since water is the bloodstream of the biosphere (Falkenmark, 2003), water management offers an opportunity for ecosystem protection that may be underestimated.

The current overshoot beyond the biocapacity of the planet is a rapidly growing dilemma that has to be addressed in order to secure a sustainable world (WWF, 2006). To secure sustainability of the life support system, the world population will evidently have to manage within the biocapacity of the planet. This concept is defined as bioproductive land area, multiplied by productivity. Both of these determinants can be improved: the former by reclamation/restoration and the latter by better management and technology. From a water perspective, it will be essential to see how water factors are involved – the current biocapacity thinking tends to be rather water blind. Focus is on land area, not on water consumed in the biological production.

Ambiguity Continues to Challenge

A basic dilemma originates from the land and water modifications underlying socio-economic development and linked to food production, timber harvesting, energy production, water supply, wastewater disposal, etc. Efforts to secure environmental sustainability have continued since the time of the Stockholm Conference in 1972, but with rather modest results. Completely different worldviews of the ecological and hydrological communities (Eulerian as opposed to Lagrangean worldviews – that is, focus on variation in a point as opposed to variation along a pathway) have delayed mutual understanding. Ambiguity in central concepts such as ecosystems and wetlands continues to delay results.

As seen from a hydrological perspective, there are two fundamental categories of ecosystems: land-based as opposed to water-based systems (Falkenmark, 2003). The former, *terrestrial ecosystems*, may be located in recharge areas in a catchment (forests, savanna grasslands), in discharge areas (springs, groundwater-fed wetlands, meadows), or be inundation dependent (flood plains). The habitat of *aquatic ecosystems* is water in rivers, lakes and deltas. Aquatic ecosystems are particularly vulnerable in the sense that their habitats integrate a whole gamut of human activities upstream, a fact that may well explain why aquatic ecosystems have been identified as the type that has suffered the largest biodiversity loss – 50% lost in the last 30 years (MA, 2005).

Why to Protect Ecosystems

During past decades, massive advocacy efforts have been directed at explaining why to protect ecosystems and biodiversity. Arguments refer to both biological and hydrological functions. In terms of the former, ecosystems are to be seen as highly beneficial by providing *ecosystem services* such as food, timber, cattle feed, climate regulation, pollination and cycling of elements (oxygen, nitrogen, carbon dioxide). They also provide *biodiversity* in terms of species richness, offering a basic immunity against exterior change from droughts, storms, pollution, climate change, etc. and thereby insurance against collapse. Species richness is thus of fundamental importance for recovery after disturbance, since species with similar functional roles can replace each other.

In terms of *hydrological functions*, the conventional 'truth' that wetlands act as sponges, reducing flood flows and releasing water during dry periods has been shown to be quite misleading (Bullock and Acreman, 2003). This misinterpretation probably originates from confusion of active as opposed to passive water storage. Many headwater wetlands in fact increase flood peaks, since by being saturated they rapidly convey rainfall to the river. Also, evaporation is more from wetlands than from other land types, reducing the flow downstream during dry periods.

Huge land and water management efforts will be required to meet the challenge of feeding humanity on an acceptable nutritional level. The land and water alterations that are generated by intensified and expanded agricultural production can be foreseen to generate impacts on water-dependent ecosystems. The need for appropriate countermeasures is therefore urgent.

Although the debate has been dominated by WHY to protect, the answers needed by the water management community are WHAT to protect and HOW to do it.

What to Protect?

In a practical situation, the water manager needs to know which are the crucial biological landscape components of importance to put focus on. What in particular to protect could be crucial ecosystem functions in the natural landscape of fundamental importance for the local life support system (Falkenmark, 2003): habitat of birds and insects; primary production of food, timber and biofuels; safe habitats for fishery; resilience in its role as insurance against ecosystem collapse, etc. It could also be essential hydrological functions that should be protected for the benefit of downstream ecological and social systems.

Since, for instance, upland forests are essential for aquifer recharge, they have to be protected to secure groundwater for populations downstream. Protection of groundwater recharge will also be essential for groundwater-dependent terrestrial ecosystems in the discharge areas such as wetlands and meadows. Upland forests also have to be protected to avoid erosion. Similarly, cloud forests in mountains have to be protected because of their role in 'harvesting' fog water and supplying it to local populations. Mangrove forests along the coast have to be protected because of their role in flood protection.

Streamflow has to be protected as it is the lifeline and water source for populations downstream, and because of its role as habitat for fish production and thereby for their food supply and income generation. Lakes of particular importance as lifelines for large populations of fishermen also have to be protected for social security reasons. Also, iconic sites in terms of specially estimated landscape components such as a beautiful local forest, a particularly beautiful lake, a wetland with especially rich biodiversity or fundamental for bird flyways have to be protected for social reasons.

One outcome of the Ramsar Convention is a list of WHATs in the sense of wetlands of special value which should therefore be protected (MA, 2005). The question is of course HOW since the water that keeps the wetland wet may be of very different origin: from drizzle raindrops for cloud forests, to rainwater for highland bogs, to groundwater seepage for groundwater-dependent biodiversity-rich wetlands, to streamflow for inundations-dependent floodplains but also for downstream aquatic ecosystems and fisheries.

How to Protect Ecosystems?

Evidently, water is a fundamental determinant to be identified in management efforts to protect ecosystems. National parks are one way to try to protect valuable ecosystems, but the water that supports or disturbs them does not respect fences. On a general level, the answer to this question is to take an *ecosystem approach* to water management (Falkenmark, 2003), in other words, paying adequate attention to hydrologic–ecological linkages and dependencies, such as between a forest and groundwater recharge made possible by its root system, or between a grazed floodplain and the periodical inundations underlying the grass production. Some of the land and water modifications degrading ecosystems are fortunately avoidable and can be minimised, whereas others are unavoidable and have to be addressed by trade-offs, based on stakeholder negotiations.

On a more specific level, the question how to protect has two dimensions: what should enter into water management and what are the institutional requirements to make it possible? The first question may be clarified through diagnostic analysis, identifying major ecological issues in a catchment, the root causes of ecosystem degradation and the causal chains involved (Duda, 2003). This analysis has to identify the water determinants, that is characteristics of certain water elements underlying the existence of a particular ecosystem. Such determinants will have to be secured through adequate water management, water quality protection or land cover protection. Strategic action plans have to be developed, incorporating for instance minimum residual streamflow for habitat protection downstream (so-called environmental flow), maximum contaminant concentrations for similar purpose, forest protection areas in recharge areas of fundamental aquifers, flood flow mimicking in a flow-control dam structure, etc.

What is now required is identification of 'bottom lines' to incorporate in multi-sector water management efforts. For aquatic ecosystems, such bottom lines may contain for instance minimum residual streamflow including seasonal variation, connectivity conditions, minimum water quality indicators or maximum salinity levels. For terrestrial ecosystems, the consumptive water use involved in biomass production from forests is an important fact that makes forestry a streamflow-impacting activity. This particular sector includes many misunderstandings, however, with scientists running behind politicians and non-governmental organisations influenced by popular debate and media highlights (Calder, 1999). For wetlands, water determinants will, as already indicated, depend on the particular kind of wetland as seen from a water perspective.

Particular efforts have to go into avoiding pitfalls in terms of remaining ambiguity in central concepts and simplistic hydrological statements. Hydrologists have to involve themselves more in clarifying relevant hydrological functions and water determinants. For the water manager, focus has to be on water–ecosystem linkages and interdependencies, clarifying what particular catchment elements to incorporate into the IWRM process. Evidently, IWRM will also have to incorporate land use, and therefore be developed into ILWRM where L stands for land use. To this end, the ecological and hydrological communities have to develop a joint meta-language, understandable by both groups.

Reasonable Control Requirements

Introduction

Water management must become proactive rather than crisis driven. It has often been stressed in the debate that the concept of water resources management itself is somewhat misleading in the sense that the challenge is also to involve and thereby manage the people depending on and making decisions about the water.

Basically, it is in the catchment that the battle for clean water and ecosystem health has to be won (SIWI, 2005); Fig. 7. Water should be seen as the common lifeblood of the basin as a whole. This will demand new thinking in the minds of people. First of all, the conviction must be clear that change is inevitable – it is part of the process of socio-economic development. Moreover, the large complexity calls for an integrated approach; the disastrous route of past management has taught us that we need to take a proactive approach; the many conflicts of interest call for balancing and a compatibility check; and we have to strive towards achieving reasonable control through laws, institutions, incentives and capacity-building activities.

Complexity

Scientists have had clear problems in addressing the man–land–water waste system as a whole, with all its physical and socio-economic interactions. It is not that scientists are unaware of the multi-dimensional character of landscapes, but that they

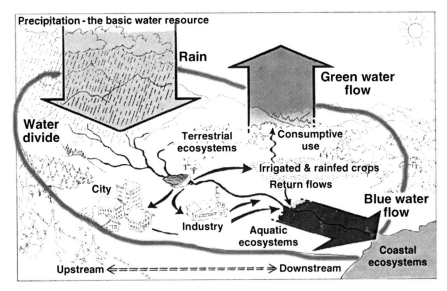

Fig. 7 Water movement through the catchment offers opportunity for integrated approach to societal water use of green and blue water and protection of ecosystems

have difficulty of practically addressing a complex and dynamic system (Falkenmark et al. 2004).

The balancing of humans and ecosystems is currently subject to rapidly growing interest in terms of needs for identifying priorities and striking of trade-offs (Falkenmark, 2003). The World Wildlife Fund's approach is to focus on particularly valuable ecosystems, identified as especially biodiverse drainage basins in a set of different freshwater ecoregions. Altogether 34 such basins had been identified in 2003 (Pittock and Holland, 2004).

As already indicated, much work has been going on for a number of years on water requirements of aquatic ecosystems in terms of the minimum residual streamflow that has to be secured. It has been suggested to be of the order of 20–30% of the average flow in countries with a definite dry season, 50% in temperate zone countries with less seasonal variations (Smakthin et al. 2004). Although current efforts have basically addressed only water quantity, water quality is probably an equally fundamental determinant behind the earlier mentioned 50% loss of biodiversity suffered since 1970. Within the European Framework Directive, a sophisticated analysis is currently ongoing on translating 'good water status' into water quality components (Heiskanen et al. 2004).

When it comes to terrestrial ecosystems, a set of widespread misconceptions have been highlighted around forests and water, in relation to water flow, dry season flow, reduction of floods, water purification, etc. (Calder, 1999). Based on field studies it has been shown that many dominating perceptions are not really true: in reality, competing processes are at work with often quite site-specific results. Some misleading perceptions might in fact be due to confusion between spatial and temporal differences: on the one hand, observations of forests being associated with particular phenomena such as relatively more rainfall, higher dry season flow and cleaner water and on the other hand effects of intentional switches in land use from one vegetation to another (deforestation, afforestation).

Compatibility

The importance of catchment dynamics originates from hydrological realities and can therefore not be neglected if foreseeable problems should be avoided (Falkenmark et al. 2004; SIWI, 2005). Basically, all water-related activities in a catchment have to be orchestrated for compatibility: water impacting land and water use, water-dependent land and water uses and ecosystems, as well as water-consumptive terrestrial and water-dependent aquatic ecosystems. Successful catchment management depends on successful involvement of all stakeholder groups, a rather difficult group to identify.

One way to move towards securing compatibility between conflicting land and water uses in a basin will, as already indicated, be through developing catchment-based IWRM into ILWRM, incorporating not only land use but also the protection of vital ecosystems, terrestrial as well as aquatic. In this process, ecologists will be crucial team members. The compatibility check also needs to focus on the rainwater partitioning process on the ground between the vertical green and the semi-horizontal blue water branches, the role of vegetation and soil in influencing such partitioning, and how the partitioning can be managed by clearing, reforestation, afforestation, etc.

There is also the evident link between downstream biodiversity and human activites upstream in the drainage basin (SIWI, 2005) to pay attention to. The habitat of aquatic ecosystems is in their water reflecting the integrated result of all human activities upstream: both upstream losses through consumptive water use, pollution loads added and agricultural chemicals leached. Measures to protect aquatic biodiversity, therefore, have to be taken upstream in the drainage basin by, for example, addressing land use and pollution load.

The way to implement a desired balancing between consumptive water use and water for aquatic ecosystems will have to be part of a catchment planning process. The process should start from the downstream end by defining ecological bottomlines in terms of the minimum residual streamflow needed and then moving section-wise upstream defining water allocation and quality conditions at each section border – an approach discussed in the Yellow River in China (SIWI, 2005).

Transboundary basins are especially challenging because of the additional dimension of interaction between sovereign governments. Governance is now being addressed by combining Track 1 activities with formal interaction between governments and Track 2 activities within civil society to secure transparency and understanding of non-negotiable natural law consequences. A particular difficulty is the task to deal with mental historic heritages in terms of hegemony (Zetoun and Warner, 2006).

A great challenge is to break intellectual and institutional barriers. Catchment area plans are an important mechanism towards a cross-sectoral approach. Basically, stakeholders need to experience the value of giving up the single sector objectives and vested interests. There has to be a 'balanced' participation of various stakeholders avoiding dominance of 'established stakeholder groups' and industry stakeholders. Events like droughts might act as external forces and help promote an intersectoral consensus. In cases where there is strong distrust in society, this might be a stumbling block in efforts to trigger the move towards catchment-based planning.

Control

Water governance is a new catchword to highlight the importance of the soft components of water resources management. It stands basically for a combination of policy and management, for example, includes processes of making choices, decisions and evaluate trade-offs, and basically covers a whole package of activities for managing the whole nexus land, water, ecosystems and society: legislation, institutions, stakeholder participation, reallocation, water banking, policy, politics, provision of water professionals, financing, incentives, sanctions, compensations, etc. It includes dialogues in which three main partners will have to be involved – government, private sector and civil society – turning the process into a trialogue. It is essential that governance also incorporate the voice of marginalised people.

Human activities on land, that is beyond the river, have to be incorporated in water-related governance efforts. This means that blue water governance also has to be expanded to incorporate green water, that is, develop into rainwater governance. Upstream–downstream linkages are especially essential, and economic compensations can turn into a useful tool in both water management and poverty alleviation.

The governance concept will have to be unpacked and disentangled to see the functions of its different components: to secure, to avoid and to foresee (Falkenmark and Rockström, 2004). *Secure* socially acceptable and environmentally feasible water supply and sanitation, food production, energy production, etc.; to *avoid* hazards from floods, droughts and bacteriological pollutants; and to *foresee* impacts such as those of adding pollution load to the water, which will reduce both the usability of the water downstream and the biodiversity there, or impacts of increased consumptive use upstream which will reduce the river flow and its dilution capacity, degrading water habitat and therefore both fish catch and biodiversity.

A proactive approach is needed to avoid water quality deterioration – streams can no longer be considered as sewer pipes. Experience has shown that pollution fees for point sources may reduce pollution loads through encouragement of best available technology. In the agricultural sector, practices are needed that reduce the discharges of nutrients and hazardous chemicals, but pollution fees may not always be appropriate. In looking for solutions for mediation of the looming world water crisis, the social and political sciences have as important a part to play as engineering and environmental science. An interdisciplinary approach is therefore imperative.

Pollution control programmes will be difficult to implement unless it is known 'who uses how much water and for what purpose, as well as who the wastewater dischargers are and what is the quality of their discharge' (Garduño, 1999). In order that water users and wastewater dischargers be willing to accept limitations, they must have legal certainty of their rights to abstract water and release wastewater. This makes water use rights administration a crucial component of water pollution abatement.

Due to the mutual stakeholder dependence, management units for river basins are needed which can work towards a long-term goal of hydrosolidarity. One such example is the river parliaments (Narain, 1999) that have been proposed in India, with its democratic tradition as a way to bring upstream and downstream stakeholders together.

Conclusions and Main Challenges

Historic Turning Point

This chapter has taken a fresh outlook towards the future, starting from today's highly fragmented environmental sector with all its different professional activities (WSS, wastewater treatment, irrigation, water storages, hydropower production, water withdrawals, out-of-basin water transfers, water pollution impacts, human health, rainfed agriculture, water management, IWRM, etc.). In a sense, the water professionals still remain ensnared in efforts to fulfil 30-year-old promises, especially the drinking water and sanitation promises originally issued by the UN Water Conference in Mar del Plata in 1977. The professionals are almost overtaken by the continuing increase of water pollution problems and are only starting to realise the risks involved in human population in general, but also that it is not impossible that water unusability may become a brake on economic development, in the sense that business moves out of polluted regions and no investors move in.

Realising the huge driving forces in terms of an additional 2–3 billion people to feed, the need to replace partly today's fossil fuels with bioenergy, and the expectations for economic development and better living conditions from billions in the developing world, implying increasing water demands, this chapter has advocated a shift in thinking and a more consistent approach. Since water is the bloodstream of the biosphere, water professionals are vital in the socio-economic development process. They will however have to find ways to link better to ecologists. Water is in fact a key entry point to ecosystem management. Moreover, it is essential to move away from remaining myths regarding water and forestry, or wetlands' water functions. In the efforts to retain reasonable control over water resources and society's water dependence in the future, it will be important to realise that:

> All science depends on its concepts, These are ideas which receives names. They determine the questions one asks, and the answers one gets. They are more fundamental than the theories which are stated in terms of them.
> G. Thomas, Nobel Prize Laureate

Three main directions have been highlighted:

- Resource adaptation sector
- Pollution abatement and technical renewal sector
- Governance sector for reasonable control

In terms of *water resource adaptation*, it has been stressed that on the one hand a number of river basins are already closed or in the process of closing so that no water remains for further appropriation and on the other hand several of them are already in the severe water shortage realm of the water scarcity diagram. With ongoing population growth they will be moving towards a situation when only the water needed for domestic and industry requirements remains. It will be essential to pay adequate attention to water constraints and linkages. This will need main shifts in water strategy and a realisation of carrying capacity limitations. An earlier neglected phenomenon has been the depletive water use linked to biomass production, the implications of which are now demonstrated by empty rivers.

Even if difficult to remedy, the *pollution abatement imperative* calls for technical renewal. Water pollution continues to increase, threatening to undermine the livelihood of the next generation.

Three sets of sources have therefore to be abated: disease agents, nutrients and hazardous chemicals, in particular persistent pollutants. The future response to the output of persistent pollutants in particular as modelled by Simonovic (2002) suggests a build-up of threats to human health, which if unaddressed might be disastrous and lead to a population catastrophe before the middle of this century. Critical action in this sector will be to develop clean production, ecosanitation, eutrophication-avoidance approaches and persistent pesticide avoidance.

The governance sector has finally to develop modes for *reasonable control*. Fundamental components in tomorrow's governance system will have to be flexible to retain reasonable control and to secure compatibility between different water-related needs and phenomena in a river basin. This will involve a remodelling of governance in order to (1) incorporate green water – a crucial water resource in many developing countries with large undernutrition today – and (2) develop an integrative capacity so that water–land ecosystems can be properly co-managed. Since it is in the catchment that conflicts of interest can be solved and trade-offs stricken, catchment-based ILWRM will be a key tool for the trade-off striking process based on stakeholder negotiations for reaching socially acceptable solutions. The governance sector will have to be flexible also to retain reasonable control over the key challenge of the future: to live with change and climate change in particular.

Core Lessons Learnt

It will be fundamental to get away from today's eddying debate highlighted by the Stockholm Water Prize Laureate 2006, Professor Biswas, in his prize lecture. Effort will be required to find out the water implications of State Leader decisions at the Millennium Summit and of reasonable expectations in developing countries for a better life. Policy makers will have to learn to see the difference between retrospective estimates and forecasting predictions. Retrospection is a valuable tool for politicians to be forewarned. What will happen if poverty and hunger are alleviated in line with the long-term goals of the MDGs?

The overall goal has to be catchment hydrosolidarity. This will call for responsible attention to ecosystems that have to be incorporated in water resources management. This will involve new expectations on the ecological community to specify bottom lines and minimum criteria that have to be protected and achieved in the process of ILWRM.

References

Agarwal A (2002) Water pollution problems posed by small industries: A case study of India and China. Proceedings of the 2001 Stockholm Water Symposium. International Water Association Publishing, London

Alcamo J, Henrichs T, Rösch T (2000) World water in 2025: Global Modelling and Scenario Analysis for the World Commission on Water for the 21st century. Kassel World Water Series Report No. 2, University of Kassel, Germany

Berndes G (2006) Future biomass energy supply: The consumptive water use perspective. Special issue: Closed basins – Highlighting a blindspot. Falkenmark M, Molden D (eds) Int J Water Resour D 24:235–245

Bullock A, Acreman M (2003) The role of wetlands in the hydrological cycle. Hydrol Earth Syst Sc 7:358–389

Calder IR (1999) The blue revolution. Land use and integrated water resources management. Earthscan, London

Colborn T, Dumanovski D, Myers JP (1996) Our stolen future. Are we threatening our fertility, intelligence and survival? – A scientific detective story. A Plume Book, Penguin Group, New York

Diamond J (2006) Collapse. How societies choose to fail or succeed. Penguin Books, New york

Duda AM (2003) Integrated management of land and water resources based on a collective approach to fragmented international conventions. Philos Trans R Soc Lond B Biol Sci 358: 2051–2062

Falkenmark M (1984) Freshwater – time for a modified approach. Ambio 15:192–200

Falkenmark M (1986) Fresh water – Time for a modified approach. Ambio 15(4):192–200

Falkenmark M (1989) The massive water scarcity now threatening Africa–why isn't it being addressed? Ambio 18:112–118

Falkenmark M (1997) Meeting water requirements of an expanding world population. Philos Trans R Soc Lond Ser B Biol Sci 352:929–936

Falkenmark M (2003) Water management and ecosystems: Living with change. GWP/TEC Background Papers No 9. Global Water Partnership, Stockholm

Falkenmark M (2005) Water usability degradation – Economist wisdom or societal madness? Invited manuscript. Water Int 30:136–146

Falkenmark M, Gottschalk L, Lundqvist J, Wouters P (2004) Towards integrated catchment management: Increasing the dialogue between scientists, policy-makers and stakeholders. Int J Water Resour Dev 20:297–309

Falkenmark M, Lannerstad M (2005) Consumptive water use to feed humanity – Curing a blindspot. Hydrol Earth Syst Sci 9:15–28

Falkenmark M, Rockström J (2004) Balancing water for humans and nature. Earthscan, London

Falkenmark M, Rockström J (2006) The new blue and green water paradigm. J. Water Resour Plng and Mgmt 132(3):129–132

FAO (1995) Land and water integration and river basin management. Proceedings of an informal workshop 31 Jan–2 Feb, 1993. Land and Water Bulletin. FAO Rome

Garduño H (1999) Water rights administration in developing countries: A prerequisite for satisfying Urban water needs. Proceedings of the 1999 Stockholm Water Symposium. Stockholm International Water Institute, Stockholm

Heiskanen AS, van de Bund W, Cardoso AC, Noges P (2004) Towards good ecological status of surface waters in Europe – Interpretation and harmonisation of the concept. Proceedings of the 2003 Stockholm Water Symposium. International Water Association Publishing, London

IWMI (2006) Insights from the Comprehensive Assessment of Water Management in Agriculture. IWMI, Colombo, Sri Lanka

Lundqvist J (1998) Avert looming hydrocide Ambio 27:428–433

MA (2005) Millennium ecosystem assessment synthesis. Island Press, Washington D.C

Narain S (1999) We All Live Downstream: Urban industrial growth and its impact on water systems. Proceedings of the 1999 Stockholm Water Symposium. Stockholm International Water Institute, Stockholm

Pittock J, Holland R (2004) Strategies to manage stream flow to benefit people and nature: a nongovernment conservation organisation's perspective. Proceedings of the 2003 Stockholm Water Symposium. International Water Association Publishing, London

Rijsberman FR (2006) Water scarcity. Fact or fiction? Agr Water Manage 80:5–22

Rockström J, Falkenmark M (2000) Semiarid crop production from a hydrological perspective. Gap between potential and actual yields. Crit Rev Plant Sci 19:319–346

Rockström J, Lannerstad M, Falkenmark M (2007) Assessing the water challenge of a new green revolution in developing countries. Proc Natl Acad Sci USA 104:6253–6260

SEI (2005) Sustainable Pathways to Attain the Millennium Development Goals – Assessint the Role of Water, Energy and Sanitation. Stockholm Environment Institute, Stockholm

Simonovic SP (2002) Global water dynamics: Issues for the 21st Century. Proceedings of the 2001 Stockholm Water Symposium. International Water Association Publishing, London

SIWI (2005) Towards Hydrosolidarity. Ample Opportunities for Human Ingenuity. Fifteen-Year Message from the Stockholm Water Symposia. Stockholm International Water Institute, Stockholm

SIWI (2007) On the Verge of a New Water Scarcity. A Call for Good Governance and Human Ingenuity. Stockholm International Water Institute, Stockholm

SIWI, IFPRI, IUCN, IWMI (2005) Let it Reign: The NewWater Paradigm for Global Food Security Final Report to CSD-13. Stockholm International Water Institute, Stockholm

Smakhtin V, Revenga C, Döll P (2004) A pilot global assessment of environmental water requirements and scarcity. Water Int 29:307–317

WWF (2006) Living Planet Report.WorldWildlife Fund. WWF International. Gland, Switzerland

Zetoun M, Warner J (2006) Hydro-hegemony – A framework for analysis of trans-boundary water conflicts. Water Policy 8:435–460

Adaptive Water Management: Strengthening Laws and Institutions to Cope with Uncertainty

Carl Bruch

Introduction

We live in a complex world full of uncertainty. This is particularly true of hydrological systems and the myriad other factors affecting water management. The nonlinear nature of the hydrologic cycle is well documented (Gleick, 1987; Lewin, 1992; Nonlinear Processes in Geophysics, 2006; Ruhl, 1997). As the debate on climate change and climate change models illustrates, it can be notoriously difficult to develop models that accurately reflect the hydrologic cycle and the factors that affect it, even when phalanxes of the world's leading scientists and computer modellers focus their attention on the task (IPCC, 2007a).

Demands on water are increasing and evolving (Varis, 2006; Gleick, 2000). Many countries have burgeoning population growth, and most are experiencing changing population dynamics as vast numbers of people move to urban centres, often clustered along coastal areas. Moreover, an overriding emphasis on economic development – and the accompanying changes in commercial and industrial demands on water that this implies – contributes to the shifting demands in water needs (Varis, 2006; Gleick, 2000). The laws, policies and institutions in place for managing water resources, however, often reflect historic uses and needs and employ outdated management approaches (Biswas, 2006; MacKay et al. 2003; Pahl-Wostl et al. 2005).

As available water resources become overdrawn and stressed, water quality problems are becoming more acute. With less water flowing, the effects of sewage, urban runoff, industrial chemicals and wastes, and agricultural fertilisers and pesticides are felt more acutely: dilution is no longer a reliable solution to pollution. While political attention from time to time focuses on the reduced quantities of available water (and the concomitant struggles over water allocation), rarely does this attention focus as keenly on water quality. Yet, it is likely that many countries will face water quality problems long before water quantity becomes a significant issue (Biswas, 2006).

C. Bruch (✉)
Environmental Law Institute,
2000 L Street NW, Suite 620, Washington, DC 20036, USA
e-mail: bruch@eli.org

These challenges are not new. For years, development planners and water managers have modelled and planned for population and economic changes in the water sector, as well as other sectors (energy, transportation, etc.). Politicians, managers and the public have become increasingly aware of – and often tried to account for – environmental considerations over the last few decades. Despite this, the laws, policies and institutions governing water and other environmental media have generally failed to keep pace with increasing demands. The Millennium Ecosystem Assessment found that approximately 60% of the ecosystem services upon which people around the world depend have been degraded or unsustainably used (MA, 2005). Over a billion people throughout the world continue to lack access to safe water sources, and 2.6 billion live without access to improved sanitation services (WHO/UNICEF, 2006).

Enter climate change. Climate change dramatically complicates the existing complexity and uncertainty that planners and managers have faced over the years. Climate change and climate variability render the entire situation more uncertain and difficult to manage. Better planning and management can enhance resilience to uncertainties in water quality and availability (this is sometimes referred to as "anticipatory adaptation") (Burton et al. 2006; Hansen et al. 2003). However, there is a growing consensus that climate change will entail significant variations in the amount and timing of precipitation, changes that are potentially dramatic or even catastrophic (Adger et al. 2007; Alley et al. 2007; IPCC, 2007a,b; Kabat and van Schaik, 2003; Arvai et al. 2006; USEPA, 2007a,b; Combes, 2003). Unfortunately, there remains significant uncertainty about the precise parameters of the location, timing and quantification regarding such changes at the regional, national and local levels. In some regions, the baseline data are so sparse that the climate change scenarios, let alone the precise implications of those scenarios, remain skeletal (Alley et al. 2007). Moreover, increasing temperatures are expected to increase water demand, for example for agriculture (Al Sairafi, 2006).

Management constraints also give rise to uncertainty. To the extent that countries have developed water laws or other environmental laws, many have yet to implement or enforce those laws effectively. So, in many instances, the feasibility and appropriateness of the current legal and institutional responses are at best uncertain. There are many reasons for the limited implementation and enforcement to date. Depending on the particular context, the reasons may include lack of financial, personnel and technical resources; lack of political will; corruption; lack of awareness by the regulators and the regulated community; and laws that are perceived as not being realistic (Zaelke et al. 2005; Kranz et al. 2005b). As a result, there often is significant uncertainty about how effective particular laws and institutions may be over the short or long term. Management is further complicated by the limited information available upon which to base decisions, as well as sensitive international politics associated with water management, particularly of shared water courses.

Despite the complexity and uncertainty inherent in water management, including as a result of climate change, there are immediate and pressing needs that must be met that require managers to make important decisions now. People need water for drinking, to grow food, for industrial and commercial purposes, and for recreation;

delaying decisions on the use and allocation of water until comprehensive studies are conducted would allow for continued unsustainable development and uses for years, further prejudicing the range of options. Similarly, it is not feasible to have a full understanding of the potential impacts of climate change before determining how best to adapt to those impacts (NRC, 2004; Hansen et al. 2003). Indeed, considering the nonlinear and complex nature of the hydrologic cycle and the socio-economic factors affecting water use (Gleick, 1987), such studies would never be able to predict accurately the long-term water needs, availability of water resources, or the specific legal and institutional responses that would be appropriate. Even if it were possible to understand and model perfectly the science of the different nonlinear physical, biological and socio-economic factors affecting water availability and use – not to mention the interplay among those factors – their sensitivity to initial conditions makes long-term predictions problematic (Gleick, 1987; Ruhl, 1996; Hornstein, 2005).

Indeed, it was attempts to model weather patterns – and the discovery of its nonlinear nature – that led to the birth of nonlinear dynamics, more popularly known as chaos theory (Gleick, 1987). One hallmark of nonlinear dynamics is that the slightest changes in physical conditions can yield different outcomes as those differences increase exponentially. Meteorologists analogised this to a butterfly flapping its wings in Hong Kong that could cause a tornado in Kansas, the so-called butterfly effect. Thus, in a nonlinear system, it is not enough to understand the laws perfectly governing the system; in order to make long-term predictions it is also necessary to have perfect knowledge about the current conditions (including what all the butterflies are doing). This is impossible. This also means that the best that one can hope for in long-term predictions are general trends.

Adaptive management provides a framework for governing water resources in a way that can account for the various uncertainties described above (Holling, 1978). Indeed, "there is broad consensus today among resource managers and academics that adaptive management is the only practical way to implement ecosystem management policy" (Ruhl, 2005). To be certain, climate change is driving much of the resurgent interest in adaptive management (consider, e.g., the dramatic growth over the last few years in the legal literature on adaptive management). However, even if climate change were not an issue, other uncertainties compel consideration of adaptive management for water resources (Neuman, 2001).

In brief, adaptive management – including adaptive water management – is an ongoing, iterative approach that seeks to "learn by doing" (Lee and Lawrence, 1986; Doremus, 2001). This includes:

- The development and adoption of a provisional legal, policy, and institutional framework
- Ongoing monitoring and collection of information
- Periodic assessment of the collected information (to determine the effectiveness of the laws and institutions)
- Modification of the legal and institutional frameworks as appropriate
- Continuing the management cycle of monitoring, assessment, and revision

Adaptive water management thus entails integrating legal and institutional approaches for managing water resources (including water allocation and pollution control), monitoring, modelling, assessing the outcomes of those management approaches, and updating and revising management decisions according to this process. As such, the legal and institutional frameworks at different levels – basin, national, local and project – need to be structured to account for the various uncertainties.

While water managers are often familiar with the principles and operation of adaptive management, most legal and institutional frameworks governing water resources have yet to incorporate specific adaptive management tools and processes. There frequently are disconnects between changing conditions and established legal and institutional responses (Ruhl, 1997): if the context is not static, why should the legal framework be static? Indeed, "stationarity is dead and should no longer serve as a central, default assumption in water-resource risk assessment and planning" (Milly et al. 2008). In many cases, legal frameworks include some of the requisite tools, including monitoring and assessment (e.g. through state-of-the-environment reports). However, these tools are generally inadequate for effective adaptive management. There is rarely any recognition in current frameworks of the necessarily provisional nature of laws and institutions, or of the associated iterative process that defines adaptive water management. Moreover, the tools that do exist tend to operate independently and are not contextualised in a broader rubric of adaptive management. What remains to be done is to put the pieces together.

This chapter examines the basic components of legal and institutional frameworks for adaptive water management. It then surveys experiences in adaptive water management from different regions and contexts. This chapter concludes with some thoughts on the future for adaptive water management, including three sets of activities to facilitate the transition to an adaptive legal and institutional framework: (1) collecting and sharing experiences to date in adaptive water management; (2) implementing pilot projects to test modalities and the extent to which approaches may be adapted to other contexts or scaled up; and (3) building capacity to develop and operationalise legal frameworks that facilitate adaptive water management.

Legal and Institutional Frameworks for Adaptive Water Management

Adaptive management recognises that there will never be complete knowledge about environmental dynamics, impacts of proposed activities, or future demands, and that many aspects will only be understood through experience, including experimentation (Walters, 1986). The lack of knowledge can be dizzying:

> ...as we know, there are known knowns; there are things we know we know. We also know there are known unknowns; that is to say we know there are some things we do not know. But there are also unknown unknowns – the ones we don't know we don't know (Seely, 2003).

Among the unknown unknowns are those things that we think we know that actually are not so. Adaptive management provides a framework for managing in light of this precarious state of (un)knowledge.

The conceptual framework for adaptive management is fairly basic, whether articulated in legal frameworks, institutional mandates or management practices. Within a particular context or problem, the initial response is developed; this may be a law, regulation, policy, permit, and so on. This response is understood to be provisional, due to the limited information that is available (Blann and Light, 2000). Once adopted, the provisional response is then implemented. This step – implementation – is the final stage for most ongoing environmental management efforts. In adaptive environmental management, however, the next step is to monitor progress towards the goals of the provisional response. Monitoring this progress entails the development and application of input and output indicators (Karkkainen, 2002; NRC, 2004; Schueller et al. 2006; Ralph and Poole, 2003) over a timeframe that is sufficient to measure the effects of a particular intervention. Periodically, managers and decision-makers take stock of the monitoring results to assess the extent to which the measures are having the intended effects, whether there are unintended effects, and whether other factors are unexpectedly affecting the outcomes. Ultimately, this periodic assessment seeks to identify whether further action is necessary (or whether the current response is adequate) as well as the nature and scope of any further action. The final stage is the modification of the initial intervention, based on the assessment; then the cycle continues with further stages of implementation, monitoring, assessment and revision.

Adaptive water management can be applied in different contexts. It may apply to the legal framework, with periodic assessment and revisions of laws and regulations governing water resource. Alternatively, it may apply within the legal and regulatory framework, for example to licenses, permits and management plans. Adaptive water management can improve long-term management of water quality and water quantity, as well as integrated management of a range of water resources. The section on "Experiences in Adaptive Water Management" surveys experiences from different contexts that may inform adaptive water management efforts.

In the three decades since Holling's seminal publication on adaptive management (Holling, 1978), the broad conceptual architecture of adaptive management has remained intact. There have been many subsequent formulations, which range along a spectrum from "active" to "passive" adaptive management (Walters, 1986; Karkkainen, 2005). Active adaptive management entails "deliberate probing for information" through experiments that test particular hypotheses. Passive adaptive management is, on the other hand, "a simpler process involving heightened monitoring of key indicators, leading to subsequent adjustments in policies in light of what may be learned through careful observation and data generation" (Karkkainen, 2005). Passive adaptive management starts with the selection of the option that is deemed the best based on existing data, and modifications are made as more information becomes available. "Evolutionary" adaptive management entails trial and error, in which random experiments and experience gradually informs understanding (Walters and Holling, 1990).

Adaptive Management and IWRM

While adaptive management is increasingly recognised as an important aspect of integrated water resources management (IWRM), it has a distinct pedigree and conceptual framework (Pahl-Wostl and Sendzimir, 2005; Medema and Jeffrey, 2005). Adaptive management was initially developed in the late 1960s by Holling and his associates, who used experimentation to understand better the dynamics and functions of specific ecosystems (Gunderson et al. 1995; Cannon, 2005). It has since evolved from use in managing land and forest resources to water resources management. In contrast, IWRM has evolved over the last century to become the dominant paradigm for water management (Rahaman and Varis, 2005).

Adaptive water management and IWRM share a number of goals and concepts, including management at the basin or watershed level and consideration of the ecological system as a whole. However, the two management paradigms are distinct. IWRM is often advocated as a politically feasible approach to managing water resources, and the practice of IWRM does not necessarily incorporate science into the management process (Pahl-Wostl and Sendzimir, 2005). Monitoring is often limited and passive, making science-based periodic assessments and revision difficult (Pahl-Wostl and Sendzimir, 2005). Moreover, while adaptive management focuses specifically on how to deal with uncertainty, IWRM generally does not explicitly focus on this issue (Pahl-Wostl and Sendzimir, 2005; Medema and Jeffrey, 2005).

Notwithstanding these differences, adaptive water management and IWRM are compatible. Indeed, to the extent that IWRM is pursued as a framework for managing a particular waterbody (Biswas, 2004), IWRM can provide a context for adaptive water management, and adaptive water management – and its emphasis on scientific monitoring, assessment and iterative revision – can serve as an essential component necessary to achieve the goals of IWRM (Kabat et al. 2003).

The Role of Law in Adaptive Water Management

Laws and regulations have a critical role to play in supporting adaptive management, for example by providing the underlying mandate for monitoring, public involvement and periodic revision (Kranz et al. 2005a). However, legal, regulatory and administrative frameworks can also hinder the application of adaptive management (Ruhl, 2005; Doremus, 2001; Coleman, 1998). In the United States, the Administrative Procedure Act (APA) and the National Environmental Policy Act (NEPA) have been interpreted and applied to focus on front-end planning while discouraging mid-course modifications. Professor Ruhl has argued that Habitat Conservation Plans (HCPs) under the U.S. Endangered Species Act (ESA) had their adaptive management features removed through procedural rules that form the basis of the APA (Ruhl, 2005), although HCPs may not have been particularly adaptive in the first place (Karkkainen, 2005). More broadly, while the ESA is conducive to adaptive management, the law is not specifically structured to enable –

let alone require – the use of adaptive management (Ruhl, 2004). Two recent U.S. cases involving NEPA further highlight the challenges of implementing adaptive management where there is no enabling legal environment defining the contours for applying adaptive management. An attempt to introduce adaptive management in the management of the Missouri River – which would help species recover but might affect navigation, power generation and flood control – was permitted as long as the adaptive management was consistent with NEPA; the Army Corps of Engineers conceded that if adaptive management led to a "major" policy change, it would undertake the time-consuming process of a supplemental environmental impact statement (In re Operation, 2005). In another case, a U.S. court invalidated a management plan for the Sawtooth National Forest that relied on adaptive management, but failed to provide sufficient details regarding its procedures, objectives or operation (Western Watersheds Project, 2006). Moreover, administrative conflicts can arise between historic top-down decision-making processes and adaptive management processes that adjust actions based on new information (Thrower, 2006). In order to provide a clear mandate for adaptive management and remove administrative, legal and regulatory barriers, Ruhl has proposed a National Adaptive Management Act, akin to NEPA since "there is good reason to doubt whether regulation by adaptive management is possible without substantial change in the administrative law system" (Ruhl, 2005).

In areas where the doctrine of prior appropriation governs the allocation of water, such as the American West, climate change threatens to cut off, rather than curtail, water consumption by junior users. In most instances, these are urban and commercial uses, which are more recent than the initial agricultural claims, although many municipalities now hold senior rights that they acquired from agricultural users. As Professor Neuman observed:

> The prior appropriation system was designed to cope with short-term shortages of the type common only in the last century. Current law is just not up to the task of handling truly severe long-term changes, such as widespread multi-decade drought, without drastic and unacceptable economic dislocations for everything from agriculture to municipal supply (Neuman, 2001).

With the very real prospect of long-term, if not permanent, changes in precipitation associated with climate change (Adger et al. 2007; Alley et al. 2007; IPCC, 2007), there is a strong need to revise legal frameworks governing water to ensure that they become more adaptive and proactive in promoting resilience to climate variability and change.

Many of the limitations in existing legal frameworks are a function of their reliance on an equilibrium-based understanding of ecosystems (Tarlock, 1994; Profeta, 1996). With the development and maturation of nonlinear dynamics and complexity theory, this understanding has been discredited (Neuman, 2001; Ruhl, 2005). At the same time, managers and regulators have experimented with new approaches, often in spite of legal frameworks. In some instances, success in incorporating adaptive management practices has led to subsequent legal or regulatory reforms, as with the development of licensing procedures for federally licensed hydropower dams in the United States (Bruch et al. 2007). Some of these experiences are discussed in the next section. These experiences demonstrate ways in which legal frameworks will need to be modified to manage actively and adaptively water resources in ways that account for new scientific understanding of ecosystem dynamics and the uncertainties wrought by climate change.

Experiences in Adaptive Water Management

Over the last few decades, there has been a growing body of experience in adaptive water management. These experiences have sought to improve management of water resources to enhance water quality and water quantity, fisheries, and consumptive and nonconsumptive uses. Most experiences to date have been in developed countries, but there are a number of initiatives underway to introduce adaptive water management in developing countries (Kranz et al. 2005a,c).

U.S. Clean Water Act

The U.S. Clean Water Act takes a phased approach to improving and maintaining water quality of the country's navigable waters (USEPA, n.d.; Gross, 2006; Houck, 2002). Initially, the act focused on reducing pollution from point sources, such as pipes. Companies and other polluters had to comply with technology-based effluent limits (with the control technologies becoming progressively more stringent) set through a national permitting system, which states could administer. It was assumed that, in most cases, such regulation would be effective in achieving the water quality goals of the act (Percival et al. 2006). In some cases, though, there may be too much effluent, not enough flow, or too much nonpoint source pollution. To provide a "safety net" for such contingencies, the act also required monitoring of ambient water quality to determine which water bodies are "water quality limited". The act then provided a process for identifying and prioritising water-quality-limited bodies (and segments of bodies), developing total maximum daily loads of specific pollutants for those water bodies, and translating those loads into permit requirements and other measures (e.g. to address nonpoint source pollution).

The 1972 Act itself has undergone a form of adaptive management through an iterative process of legislative development, implementation, monitoring, assessment, and reform and reauthorisation. As weaknesses in the act became apparent through monitoring and assessment, legislative amendments sought to address those gaps. For example, the 1977 Amendments greatly strengthened the 1972 Act; the 1981 Amendments streamlined the process for grants to build municipal treatment plants; and the Water Quality Act of 1987 inter alia phased out the grants programme and replaced it with a revolving fund. This series of amendments reflects an evolving understanding of the challenges in U.S. water management and the measures necessary to effect the desired objectives.

Murray-Darling River Basin

The Murray-Darling river basin in Australia is home to one of the largest integrated catchment management programmes in the world, covering an area of more than 1 million km^2. This programme has integrated an adaptive management approach within the broader framework of IWRM. The basin managers have recognised that decision-making "cannot wait for perfect knowledge"; rather, it is necessary to "make decisions on the best available information, and continuously improve knowledge" (Murray-Darling Basin Ministerial Council, 2001). Accordingly, the Murray-Darling programme has identified interim management targets. As the basin authority pursues those targets, it continues to monitor the status of the basin (indeed, climate change may reduce flow by 10–35% by 2050 (Pittock and Wratt, 2001)), evaluate progress in implementation every 3 years based on the monitoring information, and refine targets and implementation measures as knowledge improves. This initiative is still in the early stages: the basin-wide strategies for setting and managing targets, including the process for reviewing and revising the strategy, have yet to be developed.

The Murray-Darling experience is significant for a few reasons. First, adaptive management in the Murray-Darling basin is being implemented as a matter of policy, rather than law: there is little reference to legislation or legislative requirements. Moreover, despite the scarcity of water resources in the region (or perhaps because of the scarcity), the basin organisation is "prepared to change direction (every 3 years) on the basis of this (new) information" (Murray-Darling Basin Ministerial Council, 2001). Flexibility is not only authorised, but mandated. Finally, the geographic scale and scope of topics included in the adaptive water management scheme is extensive.

U.S. Hydropower Dam Licensing and the Clark Fork Project

Decisions regarding the licensing and relicensing of hydropower dams are often made with imperfect knowledge regarding the impacts of the dams or the effectiveness of potential mitigation measures. Since dam licences may be valid for 40 years or longer (in some countries dams are granted a permanent licence), it is particularly important that the legal and regulatory framework accounts for uncertainties in the implementation of those licences. Adaptive management provides a dynamic approach to reaching certain goals by periodically monitoring progress towards that goal and making adjustments as necessary. This necessarily entails a long-term commitment to monitoring and evaluation.

In the United States, adaptive management is being incorporated into dam licences, especially for environmental and social protection, mitigation and enhancement measures. Licences incorporate adaptive management approaches (including licence amendment procedures) to account for changes in information, technology, management practices, and context (e.g. socio-economic demands and climate change) over the term of the licence. These provisions allow licensees to adapt to changes while remaining in compliance with their licence.

The relicensing process of the Clark Fork Project in the 1990s established the Living License™, which promotes ongoing problem solving through adaptive management (Bruch et al. 2007). While the impacts of the existing dam were known, the most effective means for mitigating those impacts were unclear. An adaptive management approach provided a framework for testing different measures to see which would be the most effective and then scaling them up. Accordingly, the licensee initially implements those measures that seemed the most likely to succeed. A management committee and technical advisory committees then evaluate the effectiveness of these initial measures using the monitoring programmes established through the settlement agreement. Based on the results of these evaluations, the measures are fine-tuned or replaced, after approval by the Federal Energy Regulatory Commission (the licensing authority). A committee of signatories to the agreement – the management committee – meets biannually to monitor the operation of the dam and determine whether it is complying with the various measures of its licence. If there is noncompliance, operational changes are made with the committee's input and direction. The initial results of this adaptive management approach are promising, including improved habitat for bull trout and successful mitigation of other negative impacts of the dam.

A growing number of U.S. dams have incorporated similar adaptive management provisions into their licences (Bruch et al. 2007). These include, for example, the St. Lawrence-FDR Project (Suloway, 2006), the Pelton Round Butte Project (PGE, 2006), and Glen Canyon Dam (USBR, 2001).

Other Experiences

Managers of other aquatic ecosystems – including the Columbia River, Florida Everglades, Platte River, the North American Great Lakes, Okavango Delta, and the San Francisco Bay-Delta – have also attempted to implement adaptive management (Lee, 1993, 1999; Volkman, 2005; Farber and Freeman, 2005; Anderson and Hamman, 1996; Platte River Recovery Implementation Program, 2006; Ashton et al. 2005; Cannon, 2000; Karkkainen, 2006). Despite the availability of funding and technology, these efforts have had mixed results. For example, an initiative to manage adaptively the Columbia River devoted most of its time and resources on modelling of the river, rather than on adopting a truly iterative approach (Volkman, 2005). The restoration of the Everglades has been criticised for using a passive approach to adaptive management, in part due to the current regulatory framework (NRC, 2003; Anderson and Hamman, 1996). Similarly, the CALFED programme to manage the San Francisco Bay-Delta has suffered from monitoring gaps, for example, regarding the status of the endangered Delta smelt (Ruhl, 2004).

More productive efforts at implementing adaptive management have included the state of Oregon's response to the dramatic decline in anadromous fish in the state's waters. To mitigate this problem, the state government developed the Oregon Plan for Salmon and Watersheds (Oregon Plan n.d. a). This plan relies on active adaptive management to test hypotheses through management action, learn from the experiences, and then amend policies and management practices. Execution of this plan relies on an iterative process of planning, action, and monitoring, with strong scientific oversight by an Independent Multidisciplinary Science Team. Although the initiative is essentially non-regulatory, a series of state laws and other regulatory measures was necessary to establish it and keep it operating (Oregon Plan n.d. b).

The Netherlands national strategy for adaptation to climate change identified water as the sector that will be most affected by climate change. These effects include those related to sea level rise, river discharges, groundwater, storms and droughts (Leusink, 2006). To respond to these threats, the country is undertaking a number of legal and institutional reforms, including changes to physical planning rules so that land use follows the natural systems, as well as developing a revolving fund to support "climate proofing".

There is a rich body of experience – some effective, some problematic – from wildlife and fisheries management (Williams et al. 2007; Mitchell et al. 2005; Ruhl, 2004; Parma et al. 1998; Garaway and Arthur, 2002; Meretsky et al., 2000; Smith et al. 1998; SPU, 2005), forestry management (Thomas et al. 2006; Bormann et al. 1994; Brunner et al. 2005; Carden, 2006), land management (Doremus, 2001; Moir and Block, 2001), land conservation (Greene, 2005), aquatic ecosystem restoration (Tarlock, 2006), air quality management (Shaver, 2006) and even international trade (Cooney and Lang, 2007) that can inform the development of legal and institutional measures to manage water resources adaptively. These experiences include those in which adaptive management was mandated by law, where it was conducted within the legal framework (although not explicitly required), and examples of where adaptive management was undertaken notwithstanding gaps or impediments in the legal framework.

Status of Adaptive Management in Water Laws

Many elements of adaptive water management currently exist in the water laws and institutions of several countries; however, they usually are not contextualised in an integrated manner necessary for adaptive management. Accordingly, the existing elements may provide a framework upon which to build in adaptive water management, although they do not operate in a complementary or synergistic manner. For example, monitoring often focuses on compliance and enforcement, rather than on identifying how and why management approaches are or are not achieving their goals (although there may be some monitoring on the status of water resources). Monitoring is rarely structured to create a basis for subsequent actions. Similarly, many laws provide for periodic state-of-the-environment reports (UNEP n.d.); however, these assessments do not necessarily feed into a process for reforming the legal or policy framework. Most laws and regulations do not provide for periodic revision, the linchpin of adaptive management. Since water laws often are based on an equilibrium model of the ecosystem, policymakers engage in a two-step process of developing and implementing water law, without questioning whether the law and institutions are still appropriate and capable of achieving their policy goals, let alone gathering information to answer that question (Gleick, 2000). People are still searching for "silver bullets". What they fail to realise is that water management is undertaken in a non-equilibrium world (Farber, 2003; Wiener, 1996; Flournoy, 1996; Tarlock, 2005; Thrower, 2006).

Future Directions

Integrating adaptive management into water laws, regulations and institutions is imperative. The numerous assessments of the ongoing and potential future impacts of climate change are driving growing awareness of and emerging consensus on the need for adaptive water management. This awareness motivates calls for actions that render communities and ecosystems more resilient and capable of adapting to the effects of climate change, particularly impacts on water resources. The legal frameworks that govern water and other natural resources can impede effective adaptive management, or they can be designed to manage these resources adaptively. However, adaptive management will be effective only if laws and institutions are reformed to address the many challenges before us by incorporating mechanisms for responding to changing information and policy directions in a reasonably flexible and effective manner.

Adaptive management presents numerous challenges to conventional approaches for water resources management. In the United States, the Northwest Forest Plan was the first regional application of adaptive management in the forestry sector (Thomas et al. 2006; Bormann et al. 1994). With more than a decade of experience, this initiative offers some sober thoughts for the future development and application of adaptive management (Bormann et al. 2007). As with other adaptive management efforts – particularly those not based in specific legal requirements – "scientists are most often disappointed in what managers have been able to implement" (Bormann et al. 2007). One of the major impediments to implementation of adaptive management, and particularly testing of alternate strategies, was that "precaution trumped experimentation" as parties were reluctant to experiment unless it was known that there would be no harm. At the same time, Bormann et al. found that monitoring of data was useful in predicted and surprising ways, and managers developed a deeper appreciation of the uncertainties in long-term management of fisheries and other natural resources. Other lessons learned include the importance of framing the issues, allocating effective resources, modifying assessment approaches, difficulties in effective implementation of multiscale management, and challenges in assessing the results of adaptive management due to long timescales. Finally – and of particular importance when considering the importance of a legal mandate – "when elements of adaptive management were treated as core business ... they influenced agency decisions considerably more than elements not treated as core business" (Bormann et al. 2007).

Changing Perceptions of Change

Perhaps the greatest challenge for policymakers lies in introducing the concept of living with change (Neuman, 2001; Pahl-Wostl et al. 2005). Changing a legal framework is resource-intensive; changing it to reflect an underlying paradigm of continuous change and response may be all the more challenging. People and businesses seek stability and certainty. Adaptive management may be viewed as introducing a moving target that could undermine investments of time, money and effort. Moreover, constant flux in the legal and regulatory framework may undermine confidence in the law – how can the law command respect and compliance if it may be changed on a regular basis?

In fact, adaptive management and its attendant flexibility are not inconsistent with law and certainty (Hornstein, 2005; Wailand, 2006; Karkkainen, 2003). For example, laws can (and should) provide a normative framework for adaptive management, including clear objectives and priorities to guide water management; while the objectives remain constant, the specific means for achieving those objectives will likely evolve depending on the context, demands, and lessons learned to date. Some scholars have argued that adaptive management should also allow for the revision of specific objectives; it is possible to provide for the revision of specific objectives while maintaining the overall goals (e.g. ensuring water for human consumption and ecosystem services). The legal framework can also prescribe requirements for monitoring, provide a mandate for active adaptive management through testing of hypotheses and mandate periodic assessment of progress to guide future management interventions, including regulatory reform. It may also be possible to require a specific commitment of resources within which there is some flexibility on how to use that commitment (e.g. requiring that a specified amount of flow be dedicated to environmental uses, without specifying the particular uses). The timeframe for the review and assessment should be appropriate to the context. For example, there may be annual assessments of fisheries management to set catch limits as well as assessments every 5 or 10 years, for example, to reconsider the broader normative and management frameworks.

Concerns about change and uncertainty are understandable, but not unique to adaptive management. Businesses operate in a state of constant flux. New products come on the market, competitors challenge a business' niche and market share, there are disruptions in supply, and news about the harms or benefits of a particular product affect demand. All of these effects are largely independent of the regulatory framework. Successful businesses have become adaptive because they must. Indeed, it is not uncommon for businesses to predict dire consequences of a proposed regulatory regime, only to find after it is implemented that the costs are not nearly as serious as predicted (Ackerman, 2006). There is a natural tendency to seek to minimise uncertainty and risk, but when it is necessary, businesses, governments, and others are adaptive and rely on the basic principles of adaptive management (Ruhl, 2005; Hornstein, 2005). Making the legal and institutional frameworks adaptive, then, need not cause a revolutionary shift in how business is conducted.

Similarly, some environmentalists have expressed concern that the flexibility of adaptive management may allow agencies and regulated entities to delay or avoid taking action (Karkkainen, 2003; Thrower, 2006). The concern is that if a law establishes an adaptive framework with multiple options for pursuing broad goals, the flexibility and lack of specific standards or hard obligations could hinder enforcement of the law. This could be problematic, particularly if an agency seeks to undermine environmental protections without actually changing the law or regulations. Without clear, unambiguous requirements, efforts to challenge an agency action or inaction could be deemed to be within the allowed regulatory flexibility. Moreover, planning documents or mitigation measures (e.g. associated with an environmental impact assessment) could broadly assert that an adaptive approach will be followed without providing clear goals, measures or metrics. These are legitimate concerns, and they can be addressed through providing clear goals and metrics for assessing progress, while providing some flexibility in the precise implementation measures. Moreover, ongoing stakeholder oversight can identify and highlight potential abuses of flexibility.

By engaging a broad range of stakeholders in the process of introducing adaptive water management, governments can build awareness of the need for adaptive management and ultimate acceptance of the new legal and conceptual framework (Shindler and Cheek, 1999; Kabat and van Schaik, 2003; Karkkainen, 2006). A broadly participatory process will also enable stakeholders to provide feedback regarding how best to structure and implement adaptive management (Bruch et al. 2007).

Governance Structures

While there are many permutations of adaptive management that can be applied, depending on the particular context, the concept of adaptive management is based on a number of assumptions that may have implications for water governance structures (Pahl-Wostl et al. 2005; Quirk, 2005; Karkkainen, 2006).

Collecting, sharing and analysing data is central to adaptive management, yet transparency and information sharing can pose political challenges in many countries (van Ginkel, 2005; Karkkainen, 2002). Information is power, and there are numerous disincentives to sharing information. Moreover, while there is a great need to improve information collection about water uses and the status of water resources, monitoring entails an investment of financial and technical resources that few countries have made. Indeed, these issues are indicative of broader resource and political constraints that developing countries face in utilising scientific evidence to inform the development and amendment of environmental policy.

Stakeholder involvement is a critical aspect of adaptive management (Norton, 2005; Pahl-Wostl et al. 2005; Kranz et al. 2005c; Stiftel and Scholz, 2005; Carden, 2006). Accordingly, it is important to design governance structures that allow stakeholders to become part of the decision-making process at multiple levels (basin, subunits and locally). Stakeholder involvement can bring additional information and perspectives to bear, vet proposed decisions and the information upon which they are based, and build support for the outcome which can help in implementation (Bruch et al. 2005; Scholz and Stiftel 2005). A multi-stakeholder process that builds trust among various interests can also alleviate concerns that adaptive management can provide a loophole for individuals or organisations seeking to avoid taking action (Bruch et al. 2007). Mechanisms to ensure accountability and rule of law – including public access to courts – can also help to provide a principled means of ensuring that the law is followed while projects move forwards.

What is the appropriate scale for adaptive water governance? Adaptive management can be practised at a wide range of levels (Mitchell et al. 2005; Walters, 1986; Norton, 2005; Pahl-Wostl et al. 2005), from management of individual salmon runs to governance of international rivers. Adaptive water management similarly can happen at multiple levels, sometimes simultaneously (Karkkainen, 2006). There is often a preference to manage water adaptively at the basin level (Pahl-Wostl et al. 2005; Neuman, 2001); failure to manage on a basin level increases the vulnerability of the management to actions in other parts of the basin that are beyond the management mandate. For transboundary waters, basin-level management can present significant challenges for coordination among countries, agencies and other actors. Failure of an adaptive management plan to incorporate effectively all the subunits (political or hydrological) could contribute to gaps in monitoring and impair the overall assessment and revision processes. Moreover, since implementation measures often rely on local actions, adaptive water management needs to consider the role, effects, incentives and context of local communities (MacKay et al. 2003).

Working at the basin scale can be challenging, particularly in transboundary basins (Kranz et al. 2005a,b). It can be difficult to align basin ecosystem needs with sovereignty and politics and to determine how to incorporate diverse political and socio-economic systems into an integrated management regime (Karkkainen, 2006). The costs of developing and implementing new institutions to manage at the basin level can be significant, and the distribution of these costs across jurisdictional boundaries may be controversial (Arvai et al. 2006; Pahl-Wostl et al. 2005). The ongoing development of national and international basin organisations illustrates that these concerns can be met to varying degrees by improving coordination across jurisdictions. Inter-sectoral coordination is equally critical (Hoagland, 2005; Hanmer, 2005). For example, the Intergovernmental Panel on Climate Change noted that "Adaptation options for coastal and marine management are most effective when incorporated with policies in other areas, such as disaster mitigation plans and land use plans" (IPCC, 2001).

In order to incorporate learning from data collection and stakeholder input, institutions and decision-makers need to have the mandate, procedures and flexibility to make mid-course adjustments. Considering the various levels at which adaptive management may be practised and the ongoing decentralisation in many countries, this is a matter not only of responsive governance, but also of providing a framework for coordinating information sharing, activities, and institutions across sectors, scales, and national boundaries.

Political considerations can strongly influence the effectiveness and success of adaptive management. Political inertia can impede the development of adaptive management (Arvai et al. 2006). For example, the timescales are mismatched: politicians tend to operate on a relatively short timeframe (namely, until the next election), while adaptive management usually requires a longer timeframe (Bormann et al. 2007). Accordingly, decision-makers may not be willing to allow sufficient time to determine whether particular approaches are effective. Moreover, for transboundary waters, national political leaders are often hesitant to lose sovereignty. To build political support for governance reforms for adaptive management, it may be necessary to raise awareness about the vulnerabilities to climate change and other stressors, the difficulties in predicting water availability inherent in the nonlinear and complex hydrologic system, and security implications of failing to adopt more adaptive approaches.

Experience suggests that capacity is a central determinant of the ability to adapt (Burton et al. 2006; Adger et al. 2007; Pahl-Wostl et al. 2005). Financial, personnel, technical, information, and other resources can greatly facilitate adaptation. Since the most severe effects of climate change are projected to be in developing countries – which are least able to cope or adapt – there is a compelling need to build capacity in developing countries to adaptively manage water and other resources (Burton et al. 2006; Burton, 2000).

Transitioning to Adaptive Water Management

The transition to adaptive management will likely focus on four issues. It is first necessary to build trust. Policymakers, regulated entities and the public must become more comfortable managing with uncertainty (Cooney and Lang 2007; Clarke, 2006; Carden, 2006). This trust can be developed through carefully constructed and implemented adaptive management pilot projects at various geographic and political levels. Second, mechanisms for collecting and sharing information need to be strengthened. Most countries have such mechanisms, but they often suffer from inadequate staff, funding and technical resources. In addition, a clear legal framework for adaptive water management can provide a mandate as well as address barriers to sharing information. Third, processes need to be developed to assess periodically the information that has been gathered. The processes for collecting, sharing and assessing must be tailored to the underlying issues that need to be understood, and it is crucial to articulate these issues clearly and specifically. Moreover, it is important to determine – at the outset, if possible – how to resolve differing interpretations of the data. Finally, there needs to be an ability and willingness to revise periodically the laws, regulations, permits and other measures based on the findings of the assessments. Since provisions requiring periodic revision are not a part of most current environmental laws, especially in developing countries, policymakers need to be educated about why and how to draft the provisions.

While concern regarding the effects of climate change may drive the development of dramatic legal and institutional reforms, adaptive water management may also be introduced gradually – in an adaptive manner. A number of confidence-building measures can be undertaken without legal development or other governmental action. Such confidence-building measures can generate consensus for adaptive management, promote understanding of different constructs of adaptive management, and provide lessons learned to guide the subsequent development and implementation of adaptive management (Bormann et al. 2007). Specific measures could include dialogues on adaptive management for government officials; engaging stakeholders and other civil society members in the discussions; improving information collection, for example, through an information clearinghouse; conducting periodic assessments regarding the state of the water (similar to state-of-the-environment assessments); developing guidance, reference and training resources on adaptive management; and establishing and cultivating networks of stakeholders in the basin that are interested in adaptation and adaptive management.

Adaptive management can be advanced through official governmental ("track I") channels, through less formal channels involving professors and NGOs ("track II" channels), or both (Davidson and Montville, 1981). Track II initiatives frequently involve technical exchanges, collaboration and dialogue. Where there is political deadlock, track II initiatives can build trust, provide opportunities to learn lessons through specific approaches, and lay the groundwork for subsequent official action. For transboundary water management, the Global Environment Facility has developed a strategic assessment and planning process that relies on adaptive management and multiple confidence-building measures designed to lay the groundwork for more formal action (Bloxham et al. 2005; IAEA, 2007). Ultimately, though, formal legal approaches will be essential: "Until control and responsibility are integrated, or at least coordinated, talking about using adaptive management in allocating water is just that – talk" (Neuman, 2001).

Conclusions

There are innumerable challenges in managing water resources. Water is a multi-sectoral issue with numerous interlinkages and feedback loops: globalisation, energy, technology, information, demographic changes, and dynamic macro- and micro-economic developments all affect water demand and supply. Complexity and non-linearity in the climatic, hydrologic and social systems mean that long-term predictions are problematic. If climate change were not an issue, it may have been possible to muddle forwards as we have over the past few decades, making slow but incremental progress. Such modest progress may have been insufficient, though, in light of population growth and unsustainable patterns of development and consumption; in any case, climate change will have potentially dramatic impacts on regions around the world already facing water scarcity. Indeed, climate change is already starting to have those effects. There is no choice; we need to make our water governance systems more resilient and adaptive and we need to do it now. This will require new legal and institutional approaches. As Asit Biswas observed, "We cannot identify tomorrow's problems, let alone solve them, with today's mindsets, yesterday's knowledge, and day before yesterday's experience" (Biswas, 2006).

To date, relatively few legal and institutional frameworks have incorporated adaptive water management approaches. This must change. With the growing understanding about the potential scope, severity, rapidity and unpredictability of climate change on water resources – not to mention the impacts of population growth, demographic changes, industrial growth and other drivers of change in water use and demand – there is an urgent need to enhance the resilience and adaptability of water governance structures around the world. Even without adaptation to climate change on the policy agenda, it was evident to most that water management needed to be more resilient and adaptable; climate change makes this need manifest and urgent.

Three key activities are needed. First, we need to learn from experiences to date in adaptive water management. While different countries and regions have explored legal and policy approaches for adaptive management, there has been as yet little meaningful synthesis of experiences (Arvai et al. 2006). Targeted research should collect and analyse these experiences at different levels and identify lessons learned. This research could form the basis for empirical guidance to inform the further development of adaptive water laws and institutions. This research could also help to identify under what circumstances adaptive management is appropriate, the policy tools that can be deployed to implement adaptive management, and whether adaptive management is more effective and efficient than other forms of adaptive management (Wailand, 2006; Rothenberg, 2005). Practical guidance tools – such as legal drafter's handbooks – can then translate the lessons learned to the policymakers, decision-makers and stakeholders.

Second, pilot projects can examine the extent to which certain approaches may be adapted to other contexts or scaled up. Pilot projects can also be invaluable in testing new modalities, including streamlined adaptive management approaches that could be applied in developing countries with limited capacity. They can also help to identify specific contextual factors that affect the effectiveness of particular adaptive management approaches (Kranz et al. 2005a). For example, legislative efforts to mitigate climate change (e.g. through carbon taxes or cap-and-trade systems) have generally proceeded independent of adaptation; the growing interest in linking adaptation to climate change (e.g. by funding some adaptation efforts through carbon taxes) is still largely theoretical and pilot projects could assist in identifying viable approaches.

Third, there is widespread need to build capacity to develop and operationalise legal and institutional frameworks that facilitate adaptive water management. This includes, inter alia, data collection and data management frameworks, as well as scientific, policy and managerial capacity to manage and utilise that information. Considering the breadth of the needs, new modalities for sharing experiences and building capacity are necessary. Traditional local and regional workshops can help, but to really be effective other means are necessary. Sustainable, endogenous capacity building measures include in-country courses at universities and research centres, regional centres that can provide on-demand technical assistance, networks to share experiences and mentor, and online e-learning.

While this chapter focuses particularly on improving adaptive capacity of domestic legal and institutional frameworks governing water resources, it is also possible to introduce adaptive management into international treaties and frameworks (Feldman and Kahan, 2007), such as those governing particular basins. Indeed, one of the most successful international environmental treaty regimes – governing substances that deplete the ozone layer – has utilised an adaptive approach to monitor progress and regularly adjust the regulated chemicals and timeframes for phaseout (Hunter et al. 2006).

Adaptive management is an essential tool for societies to adapt to the effects of climate change, as well as the uncertainties and stressors inherent in dynamic social, political, economic and natural systems. It is worth noting, though, that this is but one of the necessary tools. Adaptation also entails increasing resilience, improving early warning capacity and strengthening emergency response capacities. Each of these is a separate, albeit related, topic in its own, and the question of how to improve the legal and institutional frameworks governing the respective matters merits further examination.

Water security depends on approaches that manage resources over the long-term. A group of retired U.S. generals and admirals highlighted the long-term security implications of climate change:

> Many developing countries do not have the government and social infrastructures in place to cope with the types of stressors that could be brought on by global climate change. When a government can no longer deliver services to its people, ensure domestic order, and protect the nation's borders from invasion, conditions are ripe for turmoil, extremism and terrorism to fill the vacuum (CNA, 2007).

Laws, institutions and initiatives will never be fully effective unless they account for the dynamic, complex and nonlinear character of water management. With climate change projected to become more pronounced, the need for adaptive water management will only continue to grow.

Acknowledgements The author is grateful for research assistance from Zoe Loftus-Farren, and thoughtful comments on earlier drafts by Elissa Parker, Jessica Troell, Don Anderson, William Onzivu, and Cecilia Tortajada. Any errors are those of the author.

References

Ackerman F (2006) The unbearable lightness of regulatory costs. Fordham Urban Land J 33: 1071–1096

Adger N et al. (2007) Climate Change 2007: Climate Change Impacts, Adaptation and Vulnerability. Working Group II Contribution to the Intergovernmental Panel on Climate Change Fourth Assessment Report

Alley R et al. (2007) Climate Change 2007: The Physical Science Basis, Summary for Policymakers. Contribution of Working Group I to the Fourth Assessment Report of the Intergovernmental Panel on Climate Change

Al Sairafi AB (2006) Impact of Climate Change onWater Resources, paper dated Feb 20 (Tunis)

Anderson TT, Hamman R (1996) Ecosystem management and the everglades: a legal and institutional analysis. J Land Use Environ Law 11:473

Arvai JL et al. (2006) Adaptive management of the global climate problem: bridging the gap between climate research and climate policy. Climatic Change 78(1):217–225

Ashton PJ et al. (2005) Development of a Five-Year Research Strategy for the Okavango Delta Management Plan (ODMP). Contract Report for the Okavango Delta Management Plan (ODMP) Secretariat, Maun, Botswana

Biswas AK (2004) Integrated water resources management: a reassessment. Water Int 29:248–256

Biswas AK (2006) Water management in 2020 and beyond: issues and perspectives. Presentation at international experts meeting on water management in 2020 and beyond, Zaragoza, Spain, Nov 8

Blann K, Light SS (2000) The Key Ingredients of an Adaptive Probe, available at http://www.adaptivemanagement.net/primer.htm. Accessed Feb 23 2008

Bloxham MJ et al. (2005) Training Course on the TDA/SAP Approach in the GEF International Waters Programme. University of Plymouth. 1st edn., training materials in 6 modules (Train-Sea-Coast Programme). Available at http://www.iwlearn.net/publications/misc/presentation/tdasap course 2005.zip/ view

Bormann BT et al. (1994) Adaptive Ecosystem Management in the Pacific Northwest, General Technical Report PNW-GTR-341 (Forest Service, U.S. Department of Agriculture)

Bormann BT et al. (2007) Adaptive management of forest ecosystems: did some rubber hit the road? BioScience 57(2):186–191

Bruch C et al. (eds) (2005) Public participation in the governance of international freshwater resources. University Press, Tokyo

Bruch C et al. (2007) Compendium of Relevant Practices on Improved Decision Making, Planning and Management of Dams and Their Alternatives: Compliance Theme (UNEP Dams and Development Project)

Brunner RD et al. (eds) (2005) Adaptive governance: integrating science, policy, and decision making. Columbia University Press, NewYork

Burton I (2000) Adaptation to climate change and variability in the context of sustainable development. In: Gomez Echeverri L (ed) Climate change and development, Yale School of Forestry and Environmental Studies, New Haven, USA, pp. 153–173

Burton I, Diringer E, Smith J (2006) Adaptation to Climate Change: International Policy Options. Paper for the Pew Center on Global Climate Change. Available at http://www.earthscape.org/l2/ES17447/PEW IntlPolicy.pdf. Accessed Mar 12 2007

Cannon J (2000) Choices and institutions in watershed management. William Mary Environ Law Policy Rev 25:379

Cannon JZ (2005) Adaptive management in superfund: thinking like a contaminated site. N Y Univ Environ Law J 13:561

Carden K (2006) Bridging the divide: the role of science in species conservation law. Harv Environ Law Rev 30:165–259

Clarke A (2006) Seeing clearly: making decisions under conditions of scientific controversy and incomplete and uncertain scientific information. Nat Resour J 46:571–599

CNA Corp. (2007) National Security and the Threat of Climate Change. Available at www.securityandclimate.cna.org. Accessed Apr 17, 2007

ColemanWT (1998) Legal barriers to the restoration of aquatic ecosystems and the utilization of adaptive management. Vt L Rev 23:177

Combes S (2003) Protecting freshwater ecosystems in the face of global climate change. In: Hansen LJ et al. (eds) Buying time: a user's manual for building resistance and resilience to climate change in natural systems. WWF, Washington, UDA, pp. 175–214

Cooney R, Lang ATF (2007) Taking uncertainty seriously: adaptive governance and international trade. Eur J Int Law 18:523–551

Davidson WD, Montville J (1981) Foreign policy according to freud. Foreign Policy 45:145–157

Doremus H (2001) Adaptive management, the Endangered Species Act, and the institutional challenges of "new age" environmental protection. Washburn Law J 41:50

Farber DA (2003) Probabilities behaving badly: complexity theory and environmental uncertainty. U.C. Davis Law Rev 37:145

Farber DA, Freeman J (2005) Modular environmental regulation. Duke Law J 54:795

Feldman IR, Kahan JH (2007) Preparing for the day after tomorrow: frameworks for climate change adaptation. Sustainable Dev Law Policy 8:61–69

Flournoy AC (1996) Preserving dynamic systems: wetlands, ecology and Law. Duke Environ Law Policy 7:105

Garaway C, Arthur R (2002) Adaptive Learning: Lessons from Southern Lao PDR, available at http://www.worldlakes.org/uploads/Adaptive Learning Guidelines.PDF. Accessed Apr 18 2007

Gleick J (1987) Chaos: Making a New Science. Penguin, USA

Gleick PH (2000) The changing water paradigm: a look at twenty-first century water resources development. Water Int 25(1):127–138

Greene DM (2005) Dynamic conservation easements: facing the problem of perpetuity in land conservation. Seattle Univ Law Rev 28:883–923

Gross JM (2006) Clean Water Act: Basic Practice Series. American Bar Association

Gunderson L et al. (1995) Barriers and bridges to renewal of ecosystems and institutions. Columbia University Press, NewYork

Hanmer R (2005) Chesapeake bay protection: business in the open. In: Bruch C et al. (eds) Public participation in the governance of international freshwater resources. University Press, Tokyo

Hansen LJ et al. (eds) (2003) Buying time: a user's manual for building resistance and resilience to climate change in natural systems. WWF, USA

Hoagland RA (2005) Public participation in a multijurisdictional resource recovery: lessons from the chesapeake bay program. In: Bruch C et al. (eds) Public participation in the governance of international freshwater resources. University Press, Tokyo

Holling CS (ed) (1978) Adaptive environmental assessment and management. John Wiley and Sons, New York

Hornstein DT (2005) Complexity theory, adaptation, and administrative law, Duke Law J 54: 913–960

Houck OA (2002) The Clean Water Act TMDL program: law, policy, and implementation. Environmental Law Institute, Washington, DC

Hunter D, Salzman J, Zaelke D (2006) International environmental law and policy, 3rd edn. Foundation Press, Baltimore, MD

In re Operation of the Missouri River System Litigation (2005) 421 F.3d 618 (8th Cir.)

Intergovernmental Panel on Climate Change (IPCC) (2001) Climate Change 2001: Climate Change Impacts, Adaptation and Vulnerability. Contribution of Working Group II to the Third Assessment Report of the Intergovernmental Panel on Climate Change (Geneva: UNEP/WMO)

Intergovernmental Panel on Climate Change (IPCC) (2007a) Climate Change 2007: The Physical Science Basis, Contribution of Working Group (WG) 1 to the Fourth Assessment Report of the IPCC (New York: Cambridge University Press). Available at http://www.ipcc.ch/press/index.htm

Intergovernmental Panel on Climate Change (IPCC) (2007b) Climate Change 2007: Climate Change Impacts, Adaptation, and Vulnerability, Contribution of Working Group (WG) 2 to the Fourth Assessment Report of the IPCC (New York: Cambridge University Press), available at http://www.ipcc.ch/press/index.html

International Atomic Energy Agency (IAEA) (2007) The GEF IW TDA/SAP Process: A Proposed Best Practice Approach, rev. 6. Available at http://www-naweb.iaea.org/napc/ih/document/TDA SAP/1.%20TDA-SAP%20Best%20Practice%20document.pdf. Accessed Apr 18 2007

Kabat P, van Schaik H (2003) Climate Changes the Water Rules: How Water Managers Can Cope with Today's Climate Variability and Tomorrow's Climate Change. Netherlands: Dialogue on Water and Climate

Kabat P et al. (eds) (2003) Coping with Impacts of Climate Variability and Climate Change in Water Management: A Scoping Paper. Available at http://www.waterandclimate.org/UserFiles/File/scoping.pdf. Accessed Apr 7 2007

Karkkainen BC (2002) Toward a smarter NEPA: monitoring and managing government's environmental performance. Columbia Law Rev 102:903

Karkkainen BC (2003) Adaptive ecosystem management and regulatory penalty defaults: toward a bounded pragmatism. Minn Law Rev 87:943

Karkkainen BC (2005) Panarchy and adaptive change: Around the loop and back again. Minn J Law Sci Technol 7:59–77

Karkkainen BC (2006) Managing transboundary aquatic ecosystems: Lessons from the Great lakes. Pac McGeorge Global Bus & Dev Law J 19:209–240

Kranz N et al. (2005a) Adaptive Water Management in Transboundary Contexts – A Common Research Agenda, NeWater Report Series No. 11. Available at http://www.usf.uniosnabrueck.de/projects/newater/downloads/newater_rs11.pdf. Accessed Apr 16 2007

Kranz N et al. (2005b) Transboundary Regimes in the Amudarya Basin. Berlin: Ecologic – Institute for International and European Environmental Policy. Available at http://www.ecological.de/download/projekte/1950–1999/1951/1951_del131_%20amudarya.pdf

Kranz N et al. (2005c) Governance, Institutions and Participation in the Orange-Senqu Basin, Report to the NeWater Project. Available at http://www.ecologic.de/download/projekte/1950–1999/1951/1951_del121_orangesenqu.pdf. Accessed Apr 18 2007

Lee KN (1993) Compass and gyroscope: integrating science and politics for the environment. Island Press, Washington, DC

Lee KN (1999) Appraising adaptive management, Ecol Soc 3(2):3. Available at http://www.ecologyandsociety.org/vol3/iss3/art3/

Lee KN, Lawrence J (1986) Adaptive management: Learning from the Columbia River basin fish and wildlife program. Environ Law 16:431

Leusink A (2006) Policy Making is to Foresight: Climate Adaptation in the Netherlands, Presentation at Workshop on Water Management beyond 2020 (Zaragoza, Spain)

Lewin R (1992) Complexity: life at the edge of chaos. MacMillan, New York

MacKay HM et al. (2003) Implementing the South African water policy: holding the vision while exploring an uncharted mountain. Water SA 29(4): 353–358. Available at http://www.wrc.org.za/archives/watersa%20archive/2003/october/1.pdf. Accessed Apr 17 2007

Medema W, Jeffrey P (2005) IWRM and Adaptive Management. Available at http://www.usf.uos.de/projects/newater/Deliverables/2005/NeWater%20deliverables%201st%20reporting%20period/docs/Deliverable%201.1.1_final.pdf. Accessed Apr 18 2007

Meretsky VJ et al. (2000) Balancing endangered species and ecosystems: a case study of adaptive management in Grand Canyon. Environ Manage 25:579

Millennium Ecosystem Assessment (MA) (2005) Ecosystems and human well-being 1: Current State & Trends. Available at http://www.maweb.org/en/index.aspx. Accessed Apr 7 2007

Milly PCD et al. (2008) Stationarity is dead: whither water management? Science 319:573–574

Mitchell C et al. (2005) Hawaii's Comprehensive Wildlife Conservation Strategy. Available at http://www.state.hi.us/dlnr/dofaw/cwcs/process_strategy.htm. Accessed Apr 18 2007

Moir WH, Block WM (2001) Adaptive management of public lands in the United States: commitment or rhetoric? Environ Manage 28:141

Murray-Darling Basin Ministerial Council (2001) Integrated Catchment in the Murray-Darling Basin 2001–2010: Delivering a Sustainable Future, Canberra

National Research Council (NRC), Committee on Restoration of the Greater Everglades Ecosystem (2003) Adaptive Monitoring and Assessment for the Comprehensive Everglades Restoration Plan, The National Academies Press, Washington

National Research Council (NRC) (2004) Adaptive Management for Water Resources Project Planning, The National Academies Press, Washington

Neuman JC (2001) Adaptive management: how water law needs to change. Environ Law Rep 31:11432–37

Nonlinear Processes in Geophysics (2006) Special Issue on Nonlinear Deterministic Dynamics in Hydrologic Systems: Present Activities and Future Challenges

Norton, B.G. (2005) The rebirth of environmentalism as pragmatic, adaptive management. Virginia Environ Law J 24:353–376

Oregon Plan (n.d. a) on-line at http://www.oregon-plan.org/ Accessed Apr 7 2007

Oregon Plan (n.d. b) History/Archives. Available at http://www.oregon-plan.org/OPSW/archives/archived.shtml. Accessed Apr 7 2007

Pahl-Wostl C, Sendzimir J (2005) The Relationship between IWRM and Adaptive Management, NeWater Working Paper 3. Available at http://www.usf.uni-osnabrueck.de/projects/newater/downloads/newater_rs03.pdf. Accessed Apr 4 2007

Pahl-Wostl C, et al. (2005) Transition to Adaptive Water Management; The NeWater Project. Water Policy. NeWater Working Paper X

Parma AM et al. (1998) What can adaptive management do for our fish, forests, food, and biodiversity? Integr Biol 1:16

Percival RV et al. (2006) Environmental regulation: Law, science, and policy, 5th edn. Aspen, New York

Pittock B, Wratt D (2001) Australia and New Zealand, Third Assessment Report, Intergovernmental Panel on Climate Change (IPCC)

Platte River Recovery Implementation Program (2006) Adaptive Management Plan. Available at http://www.platteriver.org/library/Program%20Document/adaptive_management_plan.pdf. Accessed Apr 18 2007

Portland General Electric (PGE) (2006) Pelton Round Butte Fact Sheet. Available at http://www.portlandgeneral.com/about_pge/news/peltonroundbutte/factsheet.asp?bhcp=1. Accessed Apr 7 2007

Profeta TH (1997) Beyond the balance of nature: environmental law faces the new ecology: managing without a balance: environmental regulation in light of ecological advances. Duke Environ Law Policy Forum 7:71

Quirk PJ (2005) Restructuring state institutions: the limits of adaptive leadership. In: Scholz JT, Stiftel B (eds) Adaptive governance and water conflict: new institutions for collaborative planning, Washington DC, pp. 204–212

Rahaman MM, Varis O (2005) Integrated water resources management: Evolution, prospects and future challenges. Sustainability Sci Pract Policy 1(1):15–21

Ralph SC, Poole GC (2003) Putting Monitoring First: Designing Accountable Ecosystem Restoration and Management Plans, In: Montgomery D, Bolton S, Booth D (eds) Restoration of Puget Sound Rivers. Available at http://www.eco-metrics.com/content/view/30/9. Accessed 23 February 2008

Rothenberg LS (2005) Incentives and adaptation, In: Scholz JT, Stiftel B (eds), Adaptive governance and water conflict: New institutions for collaborative planning, RFF Press, Washington, DC, pp 213–223

Ruhl JB (1996) Complexity theory as a paradigm for the dynamical law-and-society system: A wake-up call for legal reductionism and the modern administrative state. Duke Law J 45:849

Ruhl JB (1997) Thinking of environmental law as a complex adaptive system – How to clean up the environment by making a mess of environmental law. Houston Law Rev 34:933–1002

Ruhl JB (2004) Taking adaptive management seriously: A case study of the endangered species act, Univ Kansas Law Rev 52:1249–1284

Ruhl JB (2005) Regulation by adaptive management – Is it possible? Minn J Law Sci Technol 7:21

Scholz JT, Stiftel B (Eds) (2005) Adaptive governance and water conflict: New institutions for collaborative planning, RFF Press, Washington, DC

Schueller SK et al (2006) Evaluation Sourcebook: Measures for Progress for Ecosystem and Community-Based Projects. Available at http://www.snre.umich.edu/ecomgt/evaluation/sourcebook.html#download. Accessed 18 Apr 2007

Seattle Public Utilities (SPU) (2005) Adaptive Management Plan for Sockeye Salmon Recovery: Cedar River Watershed. Available at http://www.seattle.gov/util/About_SPU/Water_System/Habitat_Conservation_Plan–HCP/Sockeye_Hat_chery/index.asp. Accessed 18 Apr 2007

Seely H (2003) The poetry of D.H. Rumsfeld, Slate (2 Apr). Available at http://slate.msn.com/id/2081042/. Accessed 7 Apr 2007

Shaver C (2006) Air Quality Program: Information+Collaboration = Results, Natural Resource Year in Review – 2005. Available at http://www2.nature.nps.gov/yearinreview/01_B.html. Accessed 16 Apr 2007

Shindler B, Cheek KA (1999) Integrating citizens in adaptive management: A propositional analysis. Conserv Ecol 3(1):9

Smith CL et al (1998) Sailing the shoals of adaptive management: The case of salmon in the Pacific Northwest. Environ Manage 22:671

Suloway JJ (2006) Applying innovative licensing techniques: What worked at St. Lawrence. Hydro Rev (Mar) 25(1):22–27

Tarlock AD (1994) The nonequilibrium paradigm in ecology and the partial unraveling of environmental law. Loyola L.A. Law Rev 27:1121

Tarlock AD (2006) Possible lessons from a comparison of the restoration of the Danube and Colorado Deltas. Pac McGeorge Global Bus Dev Law J 19:61–78

Thomas JW et al (2006) The Northwest forest plan: Origins, components, implementation experience, and suggestions for change. Conserv Biol 20:277

Thrower J (2006) Adaptive management and NEPA: How a nonequilibrium view of ecosystems mandates flexible regulation. Ecol Law Quart 33:871–895

United Nations Environment Programme (n.d.) SoE Gateway. Available at http://www.grida.no/soe/index.htm. Accessed 17 Apr 2007

U.S. Bureau of Reclamation (USBR) (2001) Glen Canyon Dam Adaptive Management Program: Strategic Plan. Available at http://www.usbr.gov/uc/rm/amp/pdfs/sp_final.pdf. Accessed 12 Mar 2007

U.S. Environmental Protection Agency (USEPA) (n.d.) Introduction to the CleanWater Act. Available at http://www.epa.gov/watertrain/cwa/. Accessed 17 Apr 2007

U.S. Environmental Protection Agency (USEPA) (2007a) A Screening Assessment of the Potential Impacts of Climate Change on Combined Sewer Overflow (CSO) Mitigation in the Great Lakes and New England Regions, (29 Mar) Draft. Available at http://cfpub.epa.gov/ncea/cfm/recordisplay.cfm?deid=166365 Accessed 17 Apr 2007

U.S. Environmental Protection Agency (USEPA) (2007b) A Screening Assessment of the Potential Impacts of Climate Change on the Costs of Implementing Water Quality-Based Effluent Limits at Publicly-Owned Treatment Works (POTWS) in the Great Lakes Region (External Review Draft), (29 Mar) Draft. Available at http://cfpub.epa.gov/ncea/cfm/recordisplay.cfm?deid=166366. Accessed 17 Apr 2007

van Ginkel H (2005) Implications of the information society on participatory governance. In: Bruch C et al (eds) Public participation in the governance of international freshwater resources, University Press, Tokyo, pp 88–97

Varis O (2006) More urban and more aged: Demographic pressures to global water resources by 2050. Presentation at international experts meeting on water management in 2020 and beyond, Zaragoza, Spain, 8 Nov

Volkman J (2005) Sketches from life: Adaptive ecosystem management and public learning. In: Bruch C et al (eds) Public participation in the governance of international freshwater resources, University Press, Tokyo, pp 388–402

Wailand WJ (2006) Evolving strategies for twenty-first century natural resource problems. N Y Univ Law Rev 81:1518–1533

Walters C (1986) Adaptive management of renewable resources McMillan, New York

Walters C, Holling CS (1990) Large-scale management experiments and learning by doing, Ecol 71:2060

Western Watersheds Project v. U.S. Forest Service (2006) No. CV-05-189 (D. Id. 7 Feb)

WHO/UNICEF (2006) Meeting the MDG Drinking Water and Sanitation Target: The Urban and Rural Challenge of the Decade (WHO Press). Available at http://www.wssinfo.org/pdf/JMP06.pdf. Accessed 17 Apr 2007

Wiener JB (1996) Beyond the balance of nature. Duke Environ Law Policy Forum 7:1

Williams BK, Szaro RC, Shapiro CD (2007) Adaptive Management: The U.S. Department of the Interior Technical Guide. Available at http://www.doi.gov/initiatives/AdaptiveManagement/documents.html. Accessed 8 May 2007

Zaelke et al (eds) (2005) Making law work: Environmental compliance & sustainable development, vol 1. Cameron May, London

In Search of a Comprehensive Approach to Sustainable Management of Water Resources in the World Community

Kazuo Takahashi

Introduction

Water resource use has exploded in the past century and a half, and there is no end in sight in this trend. The Malthusian race between population increase and food production has heavily stressed water resources. Urbanisation has been impacting critically on water consumption, spreading rapidly to the developing world in recent decades. Industrial consumption of water resources, including indirect consumption in such a form as electricity, has become a salient feature of the modern world.

Despite all of these assaults on water, its protection has been weakening progressively. Forest coverage has been decreasing at a record pace for the past several decades. Wetlands have been filled and turned into such uses as agriculture, industry or urban dwelling. Urban areas are covered by concrete, preventing rainfall from penetrating into the soil.

The combined effects of the increasing stress on water resources and of their decreasing protection have been producing a number of grave problems. Contamination of water resources has become a major threat to human security in the world community, in particular in the developing world. Acute stress on water resources has become a major social and political issue in some countries and regions such as in northern China. On the other hand, floods have become major challenges. They are associated not only with traditional issues of control of river flows but also with such factors as hurricanes/typhoons and tsunamis. All of them take disproportionate tolls on the poor people and therefore are highly social issues. Uncertain impacts of climate change will become new issues in water resource management in less than 10 years' time.

The complex problems of management of water resources, originating in strong and enduring assault on them, in rapidly weakening protection of them, in increasingly serious rich–poor gaps, and in unpredictable but increasingly strong impacts of climate change need to be responded to with equally broad-based policy approaches. Policy responses will have to combine water resource management issues with much

---

K. Takahashi (✉)
Division of International Studies, International Christian University, 3-10-2 Osawa Mitaka, Tokyo 181–8585, Japan
e-mail: takahashi@hq.unu.edu

broader concerns. Complicated issues of policy coordination among various institutions and agencies locally, nationally and internationally will become an enormous challenge in real politics. While keeping these practical difficulties in mind, the whole process has to start with a conceptual approach. It is essential at this stage that a strong recognition of the need of a comprehensive approach to water resource management, an approach that was popular conceptually some decades ago, has to be widely shared once again and that major practical components of it should begin to be identified for implementation. In particular, in addition to traditional components, some new factors need to be added to conceptualising a comprehensive approach. The objective of this chapter is to provide a basis of discussion on these issues.

Water in the Environment

One important dimension of a comprehensive approach is to place water in the environmental context. Water itself being an essential component of the environment which is a complex whole of nature, an approach to sustainable water resource management has to be considered in the context of the natural environment.

While a number of factors in the natural environment are relevant to the issues of water resource management, forests stand out as the most important component. The stabilising function of forests in water resource provision is well known. Deforestation of upstream areas almost inevitably results in both floods and droughts in the downstream areas at the same time. These have been the major problems in many river basins around the world, including Nepal, Bangladesh and southern China.

Over the years, the importance of forests in relation to the production of oxygen, to the source of conflicts, and in recent years also to being an instrument of peacebuilding has been recognised, as well as being natural resources. In addition to these, forested areas tending to be inhabited by minority indigenous populations are politically sensitive districts. The increasing awareness of the importance of water in relation to the natural environment has added to the significance of forest management. It is important to consider the linkage between water and forests from the viewpoints of all of these inter-linked issues which tend to be neglected in dealing with water resource management.

Oxygen and Forests

The question of oxygen was highlighted at the time of the UN Conference on Human Environment in 1972. It became an issue between Brazil and the United States over the tropical forest management in the Amazon area. These forests produced 18% of oxygen around the world at that time; the North, led by the United States, applied strong pressure on Brazil for a stronger protection of these forests for the sake of the world community. Brazil countered this argument by pointing out that a large part of oxygen is consumed by industrialised countries, so this pressure is simply too egoistic and unreasonable. The debate contributed significantly 20 years later to the adoption of the principle of shared but differentiated responsibility in the Rio Declaration at the UN Summit on Environment and Development in 1992. This principle had some positive impacts on the preservation of the tropical forests in the Amazon, while the overall situation in this region continues to deteriorate significantly.

Violent Conflicts and Forests

The relation between conflicts and forest management has become an important issue since the mid-1990s, impacting significantly on water resources. The end of the Cold War brought about a large number of violent conflicts, many of them domestic, in the developing world. Research of the causes of these conflicts revealed that around 80% of them are related to conflicts over forest resources, most of which are tropical ones. Many of them are associated with conflicts between a modern legal system as adopted and implemented by the national government and customary laws that are prevailing in local communities. For example, a multinational paper company may conclude a concession agreement with the national government, including provisions of reforestation and compensation to local population, and thus assume that what it will do may even be a good model for forest resource management. As it proceeds with logging and planting, it might encounter physical attacks from the local population. It may call upon the national government to observe its part of the obligations and implement the agreement. The national government might send out a troop to protect it from attacks by the local population, resulting in armed conflicts between the troop and the local population. From the viewpoint of the local population, social relations are regulated by customary laws and the modern laws of the national government may not be clearly recognised. Therefore, the company's logging based on the concession tends to be interpreted as a wrongdoing which needs to be corrected by force if necessary. A large number of conflicts are brought about from this sort of situation.

Indigenous Population and Forests

The indigenous and minority people who live in tropical forests can often present delicate political problems in the context of national integration in many countries. Forests normally being in the marginal areas, gradual integration of them in the modern state system tends to lag behind. From the viewpoint of national leaders, it is, therefore, important that the relevant issues are dealt with without intervention from abroad. While forest management has increasingly become a global concern, the highly delicate and political nature of indigenous people makes it essential that political leaders treat it as a domestic issue. This was the major reason why developing countries, led by Malaysia, objected to the codification of forest management at the international level at the time of the Rio Summit in 1992. This issue has since been narrowed down and agreed as a set of principles, but remains politically highly sensitive.

Peacebuilding and Forests

Peacebuilding roles of tropical forests are new contributions to the recognition of the importance of forests in the world community, thus providing new impetus to improving water resource management, while they still remain at a hypothetical level to some extent. Violent conflicts took place between Moslems and Christians in Molucca, Indonesia, in 1999. This vast region around Ambon is well known for peaceful cohabitation between Moslems and Christians. Starting with the East Asian Economic Crisis of 1997, Indonesia as a whole was destabilised, leading to political crisis and independence of East Timor. Against this background, a quarrel between a young Moslem and a young Christian quickly spread to the whole region. A quarrel of two young people resulted in the mutual destruction of villages and communities, with a death toll of tens of thousands. Since 2001, various attempts at reconciliation and peacebuilding have been pursued by various actors, producing many failures and a few successes. One interesting pattern of these successes is that they are around the joint reforestation projects between Moslems and Christians. One hypothesis that appears to be proven true is related to the functions of tropical forests. The hypothesis is along the following line. Religion tends to be based on the strong sense of awe in the face of the immensity and overwhelming power of nature. Before the arrival of Islam or Christianity to Molucca, there must have been an indigenous religion which was related to the predominant power of the nature of the region. This must have been the tropical forests. Therefore, by jointly reforesting, Moslems and Christians might unconsciously be feeling the common religious traits, leading to reconciliation and healing between the Moslems and Christians. These activities are pursued by Indonesian NGOs and may have broad and significant implications to peacebuilding and reforestation activities (Takahashi, 2005).

In considering water in the environment, more specifically water in the forests, all of these factors, some of which are encouraging and others alarming, need to be considered.

Science and Technology for Water

The potential contributions of science and technology to sustainable management of water resources in the world community can be significant. They are related to the whole range of water resource issues of supply, demand and quality.

Increase in Supply of Water Resources

Since there are a number of areas where contributions of science and technology to the increase in supply of water resources can be significant, it is important to mobilise all the relevant technologies for the purpose of increasing water supply. Desalination technologies have been developed over time. While the cost factor has traditionally been the major bottleneck, the prices have been coming down to manageable levels in recent years. The major remaining problems are related to negative impacts on the environment. The enormous amounts of silts that are produced in the desalination process are environmentally hazardous wastes. Rapid increase in desalination due to decrease in costs may bring about dangerous levels of these wastes.

Recycling of used water has developed significantly. Brackish water which has been recycled is now used for agricultural purposes. With further refinement of this technology, the use of recycled water will become more common. However, one major side effect is the environmentally hazardous wastes that are produced in the process of recycling.

Technological development for the purpose of transforming these wastes from desalination and recycling to useful materials will become a major challenge in the future. Another environmental challenge is associated with transportation of water that is produced through desalination, in particular towards inland areas. Pipelines, tracks and other methods are being considered. However, all of them have significant levels of environmental damages at the moment. Due to this concern, the realistic areas where desalinated water might be provided are coastal districts where the majority of the global population lives anyway.

It is important that these and other technologies are mobilised for the production of water in a systematic manner. There is a significant techno-economic space which can increase the production of water if investment in these activities were orchestrated by resorting to both private and public sources. It is essential that environmental costs should be internalised at the planning stage.

Decreasing Use of Water Resources

Science and technology for less use of water is also an important component in a comprehensive approach to water resource management. While there are a number of technologies that have already been applied, one major area where relevant technology should contribute significantly to the decrease in water use is agriculture which is the largest sector in water usage. All of these technologies are related to control of water supply based on the application of computers. They range from large-scale irrigation to precision watering. Israel is known to have developed technologies for precision watering. These technologies may well be important diplomatic tools for Israel to attempt at a breakthrough in political isolation not only in the Middle East but also in the larger world community.

Improvement of Quality of Water Resources

Deteriorating qualities of water resources have reached crisis levels worldwide, in particular in developing countries. A major approach that has proven to be effective, which thus needs to be applied more extensively, is technology transfer. In the environment area, it is well known that the quality of the environment deteriorates rapidly with development up to a certain point where the quality begins to deteriorate less than the pace of development. With further development, the absolute levels of quality begin to improve from a certain point. This is called the environmental Kuznetzs curve. This broad progression can be applied to some environmentally damaging substances more than to others. In the industrialised countries, the first turning point took place at the level of about $11,000 per capita income (in the case of Japan in 1969), whereas it has been taking place at lower levels of income in some developing countries. For example, in the case of the Republic of Korea, this turning point took place at the per capita income levels of around $4,500 in 1987. The difference between these two income levels for a turning point is largely due to technology transfer from Japan to the Republic of Korea. This broad trend should be applicable to the water sector as well.

One concrete example has taken place in Indonesia. The Environment Centre was established in Jakarta in cooperation with Japan for the purpose of transfer of technology in measuring qualities of water and air in the 1990s. Based on the technologies acquired by Indonesian experts at this centre, the Indonesian government developed a policy of rating plants and factories on their discharge of water wastes that are measured at the emission points. Ratings from the lowest level (rate 1) to the highest level (rate 5) are announced by the government through TV and press. The reactions of people as consumers to this information are significant, impacting on the efforts of managers of plants and factories. This innovative use of transferred technologies on the part of the Indonesian government will certainly contribute to the lowering of the turning point in the environmental Kuznets curve. A systematic effort for transfer of technology to developing countries for the improvement of the quality of water resources is certainly a major area which has to be strengthened. These technologies owned by the national government, the local government or private enterprises constitute global public goods.

Comprehensive Management of Water Resources

While the importance of comprehensive management of water resources has been pointed out for some decades, and most experts agree on this point, its implementation is rare. Some top experts now discredit it as false promises. It is critically important to understand why this is the case. My assumption is that this question goes much beyond the traditional concerns of water experts and is closely related to democracy and leadership. At the same time, the largest opportunities of introduction of comprehensive management of water resources are now widely open in relation to the hottest issue in the world community, namely, peacebuilding.

Democracy and Comprehensive Management of Water Resources

Involvement of all stakeholders of water resources, such as along the river basin, means an exercise of direct democracy. The modern history has been the period when indirect democracy has been attempted in various forms such as parliamentary democracy and presidential democracy. Any form of indirect democracy being so imperfect makes us constantly concerned with various problems such as corruption, election irregularities, scandals, disinformation and involvement of underground forces. It is not surprising that the third wave of democracy from the 1980s to the 1990s as identified by Huntington has been experiencing major setbacks for the past 15 years. Many of the fragile democracies have been drawn into civil or regional conflicts, bringing some of the fragile states to failing states, and even to failed states. Some are now experiencing dictatorship originating in elections, leading to another type of failed states. Thus, introduction of representative and indirect democracy has been experiencing considerable difficulties around the world. Introduction of direct democracy being much more difficult with much less experiences than that of indirect democracy, it is no wonder that implementation of comprehensive management of water resources has continued to be a major challenge.

However, it is clear that comprehensive management approaches can be implemented only under democracy, however imperfect it is. Direct democracy is now being introduced mainly through two routes, providing us with some encouraging factors. One is through the Internet and the other is through the civil society.

Internet has made it possible for mutual communication between the elected officials and the electorate in between elections. Anyone can access the president or the prime minister through the Internet, and the president or the prime minister can respond directly to the appeal. Implications of these possibilities have not been examined thoroughly. However, it is clear that direct democracy is being introduced into indirect democracy. One major challenge for democracy in the twenty-first century is to merge between direct democracy and indirect democracy.

The emergence of civil society provides a large political space for direct democracy. An NGO's help of the elderly in a local community is directly extended to the needy. Social actions of the senior citizens, by the senior citizens, and for the senior citizens have emerged in a number of countries. A school for basic education in a slum is organised and managed by an NGO. A voluntary medical doctor may treat a patient in a remote village. All of these activities contribute to the emergence of direct democracy.

The implementation of comprehensive management of water resources being closely related to democracy, it continues to be faced with major challenges. At the same time, even in the more difficult areas of democracy, namely direct democracy, some scope for actions has emerged in recent years. Combining the Internet and civil society along the river basin, a new horizon might appear for comprehensive management of water resources.

Leadership in Comprehensive Management

Introduction of comprehensive management into a still relatively small political space of direct democracy requires astute political leadership. Putting together stakeholders who have often conflicting interests is a difficult task. The process of involving various stakeholders is a social movement which requires a clearly defined strategy and a strong capacity to implement it for mobilisation of relevant people. Management of a river basin can be effective only when it is properly placed politically in a national context. All of these tasks require a first-class political leadership. It is essential that global efforts to foster leadership in this area should be encouraged further.

An important initial step towards fostering leadership in this area is to attract a sizable number of future leaders around the world. The number will have to be of the order of thousands, given the large number of basins that need to be covered.

The second step will be to mobilise a certain number of graduate schools around the world to introduce special programmes that go beyond water issues, including public policies, environmental studies, management sciences and leadership. Textbooks covering these subjects in a comprehensive manner will have to be produced. Interactions among these graduate schools should be promoted so that they can constantly compare experiences.

The third stage will be to start a small number of pilot projects where some of the graduates of these programmes take initiatives. These experimental projects will have to be examined closely by an international team of established experts and new graduates of these programmes together.

While it takes time to groom leaders in the difficult tasks of comprehensive management, it is an essential requirement for the long-term objectives of water management. The world community as a whole will have to be mobilised for this purpose.

Peacebuilding and Comprehensive Management

Post-conflict consolidation of peace has now been conceptualised as peacebuilding. Various issues of this concept have been discussed for these 10 years intensively and extensively. They have been put together in the expert panel's report to the secretary general of the United Nations in December 2004 as peacebuilding (United Nations, 2004). It recommended the establishment of a peacebuilding commission. The then Secretary General Kofi Annan picked it up in his report titled "In Larger Freedom" in March 2005 (United Nations, 2005). The summit meeting of the United Nations adopted the proposal of the establishment of this commission in its outcome document in September 2005, and the United Nations General Assembly adopted the resolution in December 2005 for the establishment of the Peacebuilding Commission. The initial session of the Peacebuilding Commission of the United Nations took place in May 2006. In this process the interest of the world community in peacebuilding having been promoted considerably, peacebuilding has now become a centrepiece of public policy debates in the world community.

The common understanding of the importance of peacebuilding is that it is essential to go beyond the tasks of strengthening governance. This recognition is based on the fact that more than 50% of the peace accords have been broken in 5 years in the 1990s. The traditional approach having been centred around governance, new approaches have to deal with issues that are deeper than governance and to focus more on social relations. The central concern now is to rehabilitate such basic social units as families and communities. This task is conceptualised as rebuilding of social trust.

Governance combined with social trust is the new central concern of peacebuilding. Rebuilding social trust requires healing of individuals from the psychological suffering of atrocities which sometimes have been inflicted by family members such as child soldiers. Communal reconciliation is a major task without which strengthening of governance has proven to be a hollow attempt.

For the purposes of individual healing and mutual reconciliation water plays an important role. Water has traditionally been the source of divinity and the source around which a community has developed around the world. In fact, digging wells and managing them have been effective in putting the communities back to their traditional basis of relationship among villagers. It is therefore expected that water will play a critical role in the reconstruction of social trust.

It is the task of leadership to link comprehensive management of water resources to the critical roles of water in reestablishing social trust, thus peacebuilding. Democratisation being an important objective of peacebuilding, comprehensive management of water resources may contribute significantly to fostering democracy. The linkage between the two should be, and can be, with some conceptual effort reversed from democracy as a condition for the introduction of comprehensive management, to comprehensive management as a promoter of democracy.

Peacebuilding is now widely recognised as a long-term endeavour rather than special efforts of a couple of years. The linkage between rebuilding of social trust and introduction of comprehensive management of water resources will also require long-term efforts. The introduction and spreading of comprehensive management of water resources have significant opportunities in relation to peacebuilding.

Conclusions

While assault on water is continuing, protection of water is weakening; there are a wide range of new options available to achieve sustainable management of water resources. Many of these options are provided from outside the traditional concerns of the water profession. The first task of searching for a comprehensive approach to the sustainable management of water resources in the world community is to cast the net widely to cover as much potentially relevant expertise as possible. The list of expertise may be significantly long. Management of this process will require a long-term commitment and a significant amount of resources. However, given the importance of the subject at stake, the world community should spare no effort to realise this exercise.

Second, this short chapter suggests that the attempt at a broad-based approach should be conceptualised as a social movement. Various disciplines need to be mobilised to a specific objective of achieving sustainable management of water resources. Mobilisation requires leadership, organisation and a clear concept. These components need to be elaborated to make it real.

With these two bases, it is possible to develop a comprehensive approach to sustainable management of water resources on new ground in the world community. While the challenges are huge, with human survival at stake it has to be done.

References

Takahashi K (ed) (2005) Evolving concepts of peace building. International Christian University, Tokyo

United Nations (2004) A More Secure World, our Shared Responsibility, New York

United Nations (2005) In Larger Freedom; towards Development, Security and Human Rights, New York

Science, Ideology and Sustainable Development: An Actor-Oriented Approach

Peter Söderbaum

Introduction

A number of unsustainable trends concerning land, water and air can be observed locally, regionally and globally (European Environment Agency, 2005). Scientific method in a traditional sense of positivism has a role in making such observations and measurements visible. Natural science contributes to our understanding of climate change issues, water shortages of specific qualities or land degradation.

Also social sciences have a role in relation to sustainable development (SD). In addition to the focus on specific environmental and natural resource (or ecosystem) parameters, we should focus on individuals and organisations as actors. How do specific individuals as actors understand 'sustainable development' and how do they perceive their own roles in relation to SD? These kinds of questions are as relevant for water specialists as for other actors.

SD is a contested concept in the sense that there are more interpretations than one of its meaning. And the preference for one interpretation is not exclusively a matter of good science but also of subjectivity, values and ideology. No scholar can claim value neutrality in relation to present policy and development challenges. Instead, our values and ideological orientations have to be openly articulated as part of a dialogue about the future.

In what follows I will first discuss some options with respect to interpretations of SD. I will then turn to mainstream neoclassical economics and how it relates to sustainability. It is argued that universities and university scholars have a specific responsibility in opening the doors for competing theoretical perspectives in economics (and business) education and research. Rather than the present neoclassical monopoly, competition and pluralism is advocated. Among alternatives to neoclassical theory, a specific version of institutional economics will be emphasised.

Since neoclassical economics is specific in ideological terms, the present neoclassical monopoly at departments of economics cannot be defended in a demo-

---

P. Söderbaum (✉)
Department of Economics, Malardalens University,
Box 883, 72123 Vasteras, Sweden
e-mail: peter.soderbaum@mdh.se

cratic society. Competing theoretical perspectives have to be encouraged to match better the existence of different ideological perspectives in society. In addition there is a special reason to question neoclassical economics. As will be made clear in the pages to follow, neoclassical economics does not go well with the interpretation and ideology of SD as understood here.

My chapter will then focus on the role of economics education and research at universities and schools of business, its history, present position and visions for the year 2020 and beyond. Some politicians, professionals and other establishment actors might suggest that dialogue about the future of economics should be left to the university professors in economics themselves and their internal deliberations. Nothing would, however, be more unfortunate. Such an argument implies that the same establishment actors can be left without understanding the ideological role of neoclassical economics in legitimising unsustainable development trends at all levels from the local to the global.

Interpretations of Sustainable Development

As already mentioned, SD is a 'contested concept' (Connolly, 1974). There is not a single interpretation that everybody can agree upon. In social sciences, one has to live with concepts such as 'democracy', 'power', 'poverty' and 'welfare' that can be interpreted differently.[1] This is in fact not necessarily a disadvantage. New thinking is often facilitated by considering alternative ways of defining contested concepts, such as 'democracy' or 'sustainable development'.

The choice of one understanding or interpretation of SD is not just an intellectual activity aiming at clarity. A power game is going on in society where those with plenty of resources, such as transnational corporations (TNCs), see to it that their favourite interpretation is repeated at many places and arenas.

A distinction will be made between three interpretations:

1. *Business as usual.* SD stands for 'sustained economic growth' in GDP terms and 'sustained monetary profits' in business. This is the usual 'monetary reductionism' made legitimate by neoclassical economics as theoretical perspective and neo-liberalism as an ideology.
2. *Social and ecological modernisation.* Within the scope of the present political economic system, a number of minor institutional changes are possible and have been implemented in some countries. Environmental charges, environmental (or green) taxes, green labelling, environmental impact assessment, life-cycle analysis, environmental management systems, corporate social responsibility in the sense of specific codes of conduct are examples of this. Also international agreements such as the Montreal Protocol (ozone layer) and carbon trading as part of the Kyoto Protocol (climate change) can be mentioned.

---

[1] The existence of contested concepts is perhaps not limited to social sciences. As an example 'water' can be related to more than one scientific and political perspective.

3. *Readiness to consider more radical changes in present political economic systems*. Many actors such as politicians, professionals and members of civil society organisations welcome measures listed under 2 above but feel that much more is needed. They typically think in terms of what the next move should be. How can one rethink environmental and development policy in more radical terms?

Is there a 'reasonable interpretation' of SD based on the Brundtland Report (WCED, 1987) and other UN documents from the Rio de Janeiro Conference in 1992? I think so, but advocacy of one interpretation is necessarily a political act. I will suggest that SD was launched as a partly 'new' development concept. This means that 'business-as-usual' ideas are ruled out from the very beginning. This newness stands for an emphasis on:

- Multi-sector policies and multidimensional ideas of impacts
- A precautionary principle that also can be understood in security terms
- Extending horizons in ethical terms
- Democracy as central to dialogue and policy making

One-dimensional monetary (or other) thinking is no longer enough. Non-monetary impacts should be kept separate in analysis, and the idea of trading all kinds of impacts against each other (at correct prices) in monetary terms is abandoned. In any democracy there are competing ideas about appropriate values when relating one impact to other impacts. Non-monetary impacts such as pollution of water or mining of groundwater are often hard to reverse or are even irreversible, and science cannot dictate how such impacts should be traded against each other.

There are many cases where early warnings have been neglected and the precautionary principle has not been observed (Harremoës et al. 2002). The third point above can be seen as a questioning of Economic Man assumptions in neoclassical economics, according to which the consumer egoistically maximises her/his utility (from alternative bundles of commodities) while not bothering about the outcome for others. It is believed that egoism in the market place in some mysterious way will be good for all actors involved. The title of the Brundtland Report 'Our Common Future' is part of a different thinking where the individual instead is expected to extend her/his horizons to other individuals and to other kinds of biological life now and in the future. Amitai Etzioni's *I & We Paradigm* (1988) is helpful in this regard. Self-interest is always there in some form, but each individual is part of many 'we-categories' from the family or group through the local community to the nation and finally the global community. Nobody is 'other-related' in every sense, but we can all take steps in the direction of considering broader community interests. And when individuals extend their interest and concerns, they do so on the basis of their different ideological orientations.

Agenda 21 as part of the Rio de Janeiro dialogue and agreements points to a strengthening of democracy from the local to the regional and global levels as the road to SD. SD should not be dictated by the powerful corporations and nations but it is also something for grassroots civil society organisations and so on.

Interconnectedness of Science, Paradigms, Ideology and Institutions

There are many examples of cognitive and ideological blockage (Kras, 2007) in the thinking of establishment actors (politicians, professionals, etc.). At issue is how thinking and ideological priorities can be further modified and even changed radically. It is in this part that the role of science, economics and ideology comes in (Table 1) with predominant ideas in many establishment circles to the left and alternative ideas on the right-hand side.

While positivism as a theory of science tends to celebrate objectivity, value neutrality, quantitative measurement and the testing of hypothesis, recent developments in the social sciences suggest that there is also a role for subjectivism, interpretation and contextualism. Problems do not exist exclusively in the 'field' concerning land, water and air but are also connected with the mental maps, ideologies and life-styles of individuals as actors. Dialogue with influential actors about how they interpret SD may therefore be as rewarding as any other kind of study. The focus is on each actor in her or his specific context and on subjective perceptions of roles, responsibilities and so on. The dialogue is seen as a process of interactive learning for both actors involved. Sustainability is a complex challenge where few of us (if any) can claim to know the final answers.

Institutional economics (and other similar perspectives, such as social economics, humanistic economics, some versions of feminist and ecological economics) can play a role in facilitating new thinking by pointing to:

- Alternative ideas of human beings than those connected with neoclassical Economic Man assumptions
- Alternative ideas of organisations than firms maximising profits
- Alternative ideas of market relationships than neoclassical supply and demand
- Alternative ideas of welfare, decision-making and social change processes

Institutional economics will be further discussed later on in this chapter, but a few remarks are relevant at this stage. As an alternative to Economic Man assumptions,

Table 1 Predominant and alternative perspective in contemporary development dialogue

Science	Positivism	Perspectivism, including subjectivism, hermeneutics, contextualism, etc.
Paradigms in economics	Neoclassical	Institutional
Ideology	Simplistic economic growth, reductionist ideas of efficiency, preference for privatisation (neo-liberalism)	Social, Health and Environmental (SHE) aspects carefully considered as part of a holistic assessment
Institutional arrangements	e.g. World Trade Organization	A World SHE Organisation

a 'Political Economic Person' (PEP) guided by her 'ideological orientation' is proposed. The idea here is that our understanding of individuals in the economy cannot be limited to market-related roles (and even less to neoclassical ideas of the market). The individual is at the same time a citizen, professional, parent and so on. SD refers as much to the roles and responsibilities of individuals as citizens and political actors in a democratic society. Similarly, neoclassical assumptions about the profit-maximising firm will be broadened to a Political Economic Organisation (PEO) guided by its 'mission statement'. Again SD will hardly be achieved if TNCs, smaller businesses and other organisations focus exclusively on monetary performance indicators. Corporate Social Responsibility and standardised Environmental Management Systems like ISO 14 001 with connected responsibility, transparency and accountability become understandable as part of such a broader frame of reference.

In neoclassical theory, markets are understood in terms of the mechanistic forces of supply and demand. One alternative here is to leave room for social aspects of trading. Our PEP or PEO as an actor does not exclusively bother about her or his own objectives (as in neoclassical theory) but may be more or less concerned about the outcome for other actors being part of a market transaction. Here ethical concepts such as trust, fairness (as in 'fair trade'), responsibility and accountability enter once more into the picture. A holistic idea of monetary and non-monetary impacts (social, ecological, health-related, cultural) is furthermore invoked replacing simplistic monetary calculation and accounting.

This brings us to the related ideological discussion (Table 1). Neoclassical economics can be understood as a scientific perspective, but it is at the same time an ideological perspective with its specific ideas and assumptions about individuals, firms, markets and so on. An alternative to the neoclassical view, such as institutional economics, is based on a different set of assumptions and can similarly be understood as a scientific and ideological perspective.

The predominant ideology in Western societies can be described as a market ideology with monetary reductionist ideas of progress in society and business (GDP growth and profit maximisation, respectively). The consumer is free to combine her or his basket of commodities to maximise utility. Consumer preferences are assumed to be given, and the life-style of individuals is not open for discussion. These assumptions are all part of neoclassical theory. This specific market ideology (productionism, consumerism, etc.) is made legitimate through neoclassical economics.

In terms of ideology, neo-liberalism is influential these days. Neo-liberalism is often connected with the policy of Margaret Thatcher as the Prime Minister of the UK (George, 2000). Neo-liberals are sceptical to ideas about the existence of public interests and the need for public management of natural resources, such as water and land use. They argue that private property and business management of resources of all kinds, such as water, is always more efficient. Water is regarded as a commodity much like other commodities. Privatisation is then the key to a better society. These ideas have been referred to as 'market fundamentalism' and even 'the terror of neo-liberalism' (Giroux, 2004).

An alternative in terms of ideology is, for instance, SD according to the interpretations '2' and '3' above. Ecological, social, health-related and cultural aspects must be part of a 'sustainability assessment' (Söderbaum, 2006, 2007). Ethical aspects cannot be dealt with in mechanistic and technocratic terms but has to be carefully considered as part of ideological deliberations.

While SD stands for holistic thinking and analysis, neo-liberalism is built on a whole set of simplifications. In addition to neoclassical reductionism, neo-liberals want to expand the role of the market to every corner of human activity. They see potential markets everywhere and 'commodification of the economy and society' as their motto. The price and monetary average costs of delivering 1 m^3 of water become the only consideration. What happens to natural resources, fisheries, landscape, cultural artefacts, equality, employment of different groups and other ethical or ideological considerations becomes unimportant.

While articulating alternatives to neoclassical economics is important, Table 1 suggests that new thinking is needed along many lines. Neoclassical economics is closely related to positivism as a theory of science and to neo-liberalism as an ideology (Korten, 2001; Söderbaum, 2005a). Among institutions, the World Trade Organization (WTO) is made legitimate through neoclassical international trade theory and neo-liberalism. In terms of interpretations of SD, the left-hand side of Table 1 is largely compatible with '1', that is business-as-usual ideas of development, whereas the right-hand side of the table is more compatible with '2' (social and ecological modernisation) and '3' (readiness to consider radical change in political economic system) among the earlier categories of interpretations of SD. It goes without saying that the process towards new thinking, new institutional arrangements and development patterns will be facilitated if monopoly is replaced by competition and pluralism with respect to theories of science, paradigms in economics and ideological orientations.

Comparing Neoclassical and Institutional Economics

In Table 2 an attempt has been made to compare essential features of neoclassical and institutional economics, respectively. I will here assume that neoclassical micro- and macroeconomics (left-hand side of the table) is more or less known by the reader and focus on some aspects of the right-hand column. Institutional economics is often referred to as 'evolutionary economics' as in the US-based 'Association for Evolutionary Economics' (AfEE). History matters in the sense that there is inertia and path dependence in many non-monetary dimensions. Complete flexibility as in neoclassical theory seldom exists.

The concepts of PEP and PEO as actors have been explained previously. It may be further stressed that the present version of institutional economics sees the political aspect as unavoidable in approaching SD. SD is about ideology and politics, and the idea is to suggest a conceptual framework useful for the purpose of getting closer to an SD path (Söderbaum, 2000, 2008). Neoclassical economics has other

Table 2 Neoclassical and institutional economics. Overview

View of:	Neoclassical economics	Institutional economics
History	Monetary indicators may be relevant to understand the present situation	Evolutionary perspective, path dependence
Individual	Economic Man	Political Economic Person (PEP) as actor
Organisation	Profit-maximising firm	Political Economic Organisation (PEO) as actor
Market	Supply and demand	Also social and ethical aspects of market relationships
Decision-making	Optimisation	Matching of ideological orientation and expected impacts ('pattern recognition')
Approach to assessment	Cost–benefit analysis (CBA)	Positional analysis (PA) as a disaggregated and ideologically open approach
Progress in Society	GDP growth	For example SD as a multidimensional concept
Social and institutional change	Competition, changes in supply and demand of commodities	Also debate about ideology and paradigm. Actor-institutional change processes

political purposes, for instance focusing on the fields of macroeconomic monetary policy, financial policy and the like.

Also market relationships have been touched upon earlier in this chapter. The market may be understood in mechanistic terms, but behind market transactions are PEPs and PEOs as actors. They may use their power to 'exploit' each other or engage in 'fair trade'. There is a social element, for instance, in business-to-business relationships where trust, goodwill and other long-run considerations enter into the picture (Ford, 1990).

There is a long tradition of mathematical optimisation in neoclassical theory. Monetary profits are maximised subject to various constraints. A different way to approach decision-making is to think in terms of 'matching', 'appropriateness' and 'pattern recognition'. Quantitative measurement is part of this approach but also are qualitative and visual aspects. It is assumed that each actor is guided by an 'ideological orientation' that may be fragmentary and uncertain but nevertheless points in specific directions. In a decision situation this ideological orientation is matched against the expected impact profile of each alternative considered. The alternative which best matches the ideological orientation of our PEP or PEO is chosen or at least regarded the best as part of a continued search process.

At the societal level, neoclassical economics recommends cost–benefit analysis (CBA) as the optimisation technique. It is argued that politicians and other decision-makers need a clear-cut answer to the problem faced and that money is the natural measuring rod. Some impacts of building a road or channel to transfer water are monetary in kind, such as construction costs. For other impacts the analyst refers to estimates of a 'correct' market price (at the time of study). Monetary costs and benefits are then added to a present value using a 'correct' discount rate. The analysis is technocratic and does not leave much room for public debate and ethical/ideological considerations.

CBA has been criticised over the years (e.g. World Commission on Dams, 2000; Ackerman and Heinzerling, 2004; Söderbaum, 2006), but neoclassical economists are not willing to consider alternative approaches seriously. The World Commission on Dams advocates a specific multiple criteria approach to decision-making. Positional analysis (PA) is one alternative that is transdisciplinary and ethically/ideologically open rather than closed. This approach has been described elsewhere (Söderbaum, 2000, 2005b; Bebbington et al. 2007) but there are other approaches as well claiming usefulness in relation to what is now referred to as 'sustainability assessment'.

Gross domestic product (GDP) is an essential part of neoclassical macroeconomics, and this idea of progress in society is made legitimate through neoclassical theory. GDP may be a relevant indicator for some purposes, but 'monetary reductionism' can hardly be seen as relevant in relation to SD. For the latter purpose we need a large number of non-monetary indicators. When facing complexity one has to live with some part of that complexity rather than assume it away.

Our last item in Table 2 is about social and institutional change processes. The right-hand side in a sense brings together a large part of the arguments in the present chapter. At the heart of the processes is the individual as actor and the models he or she uses in interpreting the world. Cognitive as well as value elements are part of the ideological orientations of PEPs and PEOs. The models or interpretations referred to can partly be chosen by actors:

> By deliberately changing the internal image of reality, people can change the world. (Willis Harman in Korten, 2001, p. 233)

An actor's interpretation of a phenomenon may be manifested, for instance in:

- The name given to the phenomenon
- Its acceptance among other actors
- Professional and other organisational changes
- New rules, for instance national and EU regulations

There will normally be competing interpretations, for instance those that are well established as part of the neoclassical paradigm: 'A business firm is an entity that maximises profits or shareholder value'. At some development stage, some actors within and outside business corporations understand that business operations are not necessarily innocent in environmental terms. Actors in a business company may wish to measure environmental performance since this is part of the public agenda. In addition to monetary performance, environmental management systems (EMS) such as EMAS in the European Union or ISO 14 001 appear on the scene. This then becomes a step towards multidimensional measurement of business performance. At issue is if other actors will embrace the idea of EMS or stick to a more neoclassical understanding of the business firm. Cultural differences exist, for instance between northern Europe (where EMS are relatively popular) and the United States where a reluctance to move away from traditional ideas can be observed.

Sustainability Policy in a Regional Perspective

Sustainability policy is a concern for nations and the European Union. But this is not enough. In principle, sustainability is a concern for every individual and every organisation as PEP or PEO. It is also a concern for each local and regional community. Aragon in Spain and Mälardalen in Sweden exemplify regions where different sustainability policies and projects can be considered. An intensified public debate is needed where global concerns and future generations play a role.

Politicians in the Mälardalen (or Aragon) region cannot promise their citizens that sustainability issues will be solved through policies within the region. In many cases problems are interregional and international. Climate change is a global issue that calls for efforts of a new kind. In addition to internal policy, there is therefore a need for external policy of cooperation. Networks of regions inside and outside the European Union may strengthen their efforts of CO_2 reduction.

The ideas about economics among politicians and journalists in the European Union are fairly established and closely related to neoclassical economics. Monetary reductionist ideas about progress in society and business are not compatible with a policy for sustainability. For key actors such as former Trade Commissioner Peter Mandelson in the European Union, reducing tariffs, according to the WTO philosophy is always a step in the right direction. Mandelson relied on a narrow international trade theory trying to reduce the average monetary costs of producing specific commodities such as shoes, overlooking qualitative differences relating to fashion, for instance, and all other impacts that have been discussed here. A more holistic sustainability assessment might point in other directions. Shoes are a matter of culture, and why should all kinds of shoe production be eliminated in Italy and other parts of Europe? It can be argued that we should take care of environmental impacts from such industries and not export environmental degradation to other countries. When Mandelson and the Swedish Trade Minister at the time Thomas Östros argued in favour of the so-called free trade, they probably did it because they did not know about alternatives to neoclassical trade theory.

This suggests that the neoclassical monopoly at departments of economics is one of the problems that have to be faced. Mandelson and Östros could of course argue for their position in a democratic society. But universities have no right to protect the neoclassical paradigm, thereby excluding other perspectives. Economics education has to become pluralistic.

This chapter does not primarily address issues at the ecosystem level. But it is clear that land-based ecosystems and water ecosystems should be protected in various ways. In a European Union context 'urban areas and infrastructure increased by more than 800 000 ha between 1990 and 2000, a 5.4% increase and equivalent to the consumption of 0.25% of the combined area of agriculture, forest and natural land' (European Environment Agency, 2005, p. 42). This is alarming in relation to the climate change issue and also tells us something about how we relate to non-human forms of life. Ecological footprints within and outside the European Union have to be seriously considered, including questions of neoclassical ideas of efficiency. Where water is concerned, a move away from supply management to demand management is called for involving some change in agricultural and industrial practices and use patterns as well as changes in life-styles.

Proposals of the above kind are of course always open to debate. My main recommendation, however, is to use holistic methods for sustainability assessment to prepare decisions related to land use and water use, for instance those related to construction of water transfer projects and dams.

Towards Pluralism in Economics

Many actors feel that it is unnecessary to discuss abstract theories in relation to water problems and land-use problems. They want to go directly from observations in the field to policy measures and action. This is certainly understandable, but the theoretical perspectives and mental maps of key actors and all other actors also deserve attention. The reasons have already been explained. Each theoretical perspective, particularly in economics and other social sciences, is 'scientific' as well as 'ideological'. In a democratic society, university departments of economics cannot therefore limit research work and education to the neoclassical paradigm.

An enquiry into the history of economic ideas suggests that there have always been tensions between paradigms, one being dominant at a certain time but normally challenged by competing ideas (Fusfeld, 1994). Although denied by neoclassical economists, this is also the situation today. Institutional theory appears to be strong among competing perspectives, for instance at some departments of economic history. But research in fields such as socio-economics, social economics, ecological economics and feminist economics is not limited to institutional theory.

Among recent developments, it should be mentioned here that students of economics at respected universities in France and England have questioned the kind of education offered and textbooks used at departments of economics. They are critical to the emphasis on mathematical models and the rather naïve use of mathematics in the form of equations in introductory and more advanced texts. A petition by students initiated an intervention by the Minister of Education in France. Students and some of their teachers have joined forces in a Post-Autistic Economics network, which later became a *Post-Autistic Economics Review*, an Internet-based newsletter (www.paecon.net). Three books edited by Edward Fullbrook have followed (2003, 2004, 2007), where a number of professors have contributed.

Signs exist that politicians and civil servants are also beginning to ask for something other than neoclassical theory. It is increasingly understood that it is an illusion that science can be easily separated from politics and in 2003, the German Ministry of Education and Research turned to one of the more respected German economics research institutes, DIW (Deutsches Institut für Wirtschaftsforschung), asking for a 'sustainability economics' as opposed to neoclassical economics based on the judgement that the latter is 'not adequate' in relation to some of the most important questions of our time (Dröge, 2003).

An International Confederation of Associations for Pluralism in Economics (ICAPE) has been formed (www.icape.org), and textbooks are being written that respond to the demand expressed by students and other actors (e.g. Söderbaum, 2008). Hopefully, this will lead to new options for students. But again, actors other than professional economists have to take part and express their opinions. Neoclassical economists have been successful in protecting their monopoly for a long time, for instance through the Bank of Sweden's Prize in Economic Sciences in memory of Alfred Nobel. So far this award has legitimised mainstream neoclassical economics and has not done much to move us closer to a 'sustainability economics'.

References

Ackerman F, Heinzerling L (2004) Priceless. On knowing the price of everything and the value of nothing. The New Press, New York

Bebbington J, Brown J, Frame B (2007) Accounting technologies and sustainability assessment models. Ecol Econ 61(2/3):224–236

Connolly WE 1993 (1974) The Terms of political discourse, 3rd edn. Blackwell, Oxford

Dröge S (2003) International institutions for sustainability. DIW, Deutsches Institut für Wirtschaftsforschung (German Institute for Economics Research). Research Notes 30. Berlin (www.sustainabilityeconomics.de)

Etzioni A (1988) The moral dimension. Towards a new economics. Free Press, New York

European Environment Agency (2005) The European Environment. State and Outlook 2005. Copenhagen. http://reports.eea.eu.int/stateofenvironmentreport2005/

Ford D (ed) (1990) Understanding business markets. Interaction, relationships, networks. Academic Press, London

Fullbrook E (ed) (2003) The Crisis in economics. The post-autistic economics movement: the first 600 days. Routledge, London

Fullbrook E (ed) (2004) A guide to what's wrong with economics. Anthem Press, London

Fullbrook E (ed) (2007) Real world economics. A post-autistic economics reader. Anthem Press, London

Fusfeld DR (1994) The age of the economist, 7th edn. HarperCollins, New York

George S (2000) A short history of neoliberalism. Twenty years of elite economics and emerging opportunities for structural change. In: Bello W, Bullard N, Malhotra K (eds) Global finance. New thinking on regulating speculative capital markets, Zed Books, London, pp 27–35

Giroux HA (2004) The terror of neoliberalism. Authoritarianism and the eclipse of democracy. Paradigm Publishers, Boulder

Harremoës P, Gee D, MacGarvin M, Stirling A, Keys J, Wynne B, Guedes Vaz S (eds) (2002) The precautionary principle in the 20th Century. Late lessons from early warnings. (European Environmental Agency). Earthscan, London

International Confederation of Associations for Pluralism in Economics (www.icape.org)
Korten DC (2001) When corporations rule the world, 2nd edn. Kumarian Press, Bloomfield, Conn
Kras E (2007) The blockage. Rethinking organizational principles for the 21st century. American Literary Press, Baltimore, Maryland
Post-autistic economics review, now Real-world Economics Review (www.paecon.net)
Söderbaum P (2000) Ecological economics. A political economics approach to environment and development. Earthscan, London
Söderbaum P (2005a) Actors, problem perceptions, strategies for sustainable development. Water policy in relation to paradigms, ideologies, institutions. In: Biswas AK, Tortajada C (eds) Appraising sustainable development. Water management and environmental challenges, Oxford University Press, Oxford, pp 81–111
Söderbaum P (2005b) Democracy, decision-making, and sustainable development: damconstruction as an example. Int J Water 3(2):107–120
Söderbaum P (2006) Democracy and sustainable development. What is the alternative to CBA? Integr Environ Assess Manag 2(2):182–190, April
Söderbaum P (2007) Issues of paradigm, ideology and democracy in sustainability assessment. Ecol Econ 60:613–626
Söderbaum P (2008) Understanding sustainability economics. Towards pluralism in economics. Earthscan, London
The World Commission on Dams (The Report by the World Commission on Dams) (2000) Dams and development. A new framework for decision-making. Earthscan, London
The World Commission on environment and Development (The Brundtland Report) (1987) Our common future. Oxford University Press, Oxford

Leading and Managing Change in Water Reform Processes Capacity Building Through Human Resource Development

Hans Pfeifer

Introduction

The chapter sets out by describing the highly dynamic and complex environment in which organisations in the water sector have to operate today.

Section 2 shows the dynamic and consequent change as a persistent paradigm, which is in need of leadership. Leadership sets the necessary framework to adapt to these changes at an organisational level and yet remain faithful to one's values. Steering change processes become, therefore, a core competence of leadership.

Section 3 takes us on a journey into practical experience. The examples of a national and a regional water agency show typical patterns of reaction to change. It shows gaps in management and leadership that can be addressed by capacity building.

Section 4 argues for the need to combine personnel, organisational and system development.

Section 5 describes the InWEnt capacity building concept that builds on these three levels of intervention.

Organisations in the Water Sector Today

To mention the good things first, organisations in the water sector are often highly professional bodies, following high professional values and doing a very efficient job where they have their core competences, that is, in their respective technical fields. However, the technical challenges are more and more paired with managerial ones, for which the professional body in the water sector is in considerable need of capacity building.

Organisations in the water sector in a given country exist in multi-faceted and multi-level institutional arrangements. They operate at national, regional and local levels. They are influenced by neighbouring countries, regions or localities as well as

---

H. Pfeifer (Late)
InWEnt, Department of Environment, Wielinger Strasse 52, 82340 Feldafing, Germany

by international bodies. They have functional differentiations and links, for example between regulatory bodies and drinking water providers. They are influenced by interests in other sectors such as agriculture, industry or tourism (to name but a few). They are equally influenced by political interests at all levels of administration and maintain respective links to them.

Furthermore, a great variety of challenges have to be addressed in complementarity with other systems. Examples are good governance with regard to resource management, the management of the consequences of urbanisation and simultaneous ruralisation, the increasing pollution and decreasing quality of the resource, the requirements of food security and health, crisis prevention and management, and changes in demographic patterns and of course globalisation.

This situation of embeddedness implies that organisations in the water sector need to respond to a complex and constantly fast-changing environment. Management in the sector, therefore, is far from being limited to technical manoeuvres. However, not only is the environment highly dynamic, but also the water sector organisations in a given country are undergoing massive changes. A water directorate is moved from one ministry to another and split across two departments. Decentralisation assigns new tasks and roles to old organisations. New subsidiary organisations are created. Organisational structures and practices change in response to new paradigms such as Integrated Water Resource Management. Privatisation brings about new roles along with ethical questions. In this change, organisations carry the burden of their past. They tend to remain:

- Bureaucratically and hierarchically organised (overregulated)
- Overstaffed and underpaid resulting in a lack in drive and motivation
- Technically oriented, resulting in inertia, where change management would be needed
- Governed by poor decision-making and contradicting priorities
- Politically strangulated and financially under-funded

Nevertheless, the capacities to deal with the complex and changing sector and its environment are little developed. Organisations in the water sector are frequently headed by engineers or former technicians, whose entry card into managerial positions was technical performance. Following the 'Peter Principle', they rise in the hierarchy to a point where they are overtaxed by managerial tasks and spend the rest of their professional careers hiding that shortfall. Civil Service personnel are not experienced or trained for entrepreneurship in autonomous or semi-autonomous bodies and half-heartedly follow the shift towards autonomy and a client orientation with unclear roles and massive know-how gaps. As a consequence, in most cases, the modus operandi continues to follow a technical orientation, in which managers constantly deal with emerging urgencies and feel as victims of higher authorities. Following the principle 'if my only tool is a hammer, all problems become nails', solutions are often sought in 'the home garden' of technical innovations. For instance, Geographic Information Systems (GIS) have become a technical option and hence state of the art. The technically oriented mind is quickly tempted to use GIS to structure and steer the complexity of data and may remain busy with that endeavour for 2 or 3 years, without noticing that decisions are not improving, that resources are bound and that the very existence of the GIS now requires its continuation. A typical conversation on this issue sounds like the following:

'Why did your organisation choose GIS for your data management?' 'Well, everybody does it'. 'What do you need all this data for?' 'Well, it's there and we can have it'. 'Has this improved the quality of decisions?' 'Well, it is not for me to judge'.

Thus, sometimes technical solutions for yesterday's problems become today's technical or managerial problems,

Even though this pattern becomes quite obvious, little is done to address the challenge of missing capacities: 'We are so busy at the operational level that we have no time to manage the change'.

We will illustrate that issue with a capacity building diary later in this chapter.

Change as a Persistent Paradigm and the Need for Leadership

In current development and globalisation trends, change is the only stable pattern. The ability to respond to change will decide whether we can successfully champion the development of the water sector. Change will also impact the future and the survival of organisations in the water sector. It is the skilful 'cha-cha-cha' of recognising and addressing *cha*nge, *cha*llenges and *cha*nces proactively that must be added to the world of technical solutions.

The level of action is always embedded in organisations made up of human beings and not in a faceless 'sector' or 'tool'. Therefore, we need to look at what makes organisations perform and how we can remove stumbling blocks so as to access the enthusiasm and motivation of the professionals working in them. In this context, leadership plays a key role. Even if solutions may be necessary at the level of a technical process, it is the leadership that provides the orientation and the mechanisms to plan, run, evaluate and adapt the process continually in order to make best use of its potential (just as in the 'plan-do-check-act' circle in the Deming model of quality management).

In practically all management and quality management models, you find the leadership in the centre of the organisation. According to the EFQM model (European Foundation for Quality Management), leadership is defined as follows:

> Excellent leaders develop and facilitate the achievement of the mission and vision. They develop organisational values and systems required for sustainable success and implement these via their actions and behaviours. During periods of change they retain a constancy of purpose. Where required, such leaders are able to change the direction of the organisation and inspire others to follow.

Five sub-criteria further define elements that should be addressed by leadership:

1. Leaders develop the mission, vision, values and ethics and are role models of a culture of excellence
2. Leaders are personally involved in ensuring the organisation's management system is developed, implemented and continuously improved
3. Leaders interact with customers, partners and representatives of society
4. Leaders reinforce a culture of excellence with the organisation's people
5. Leaders identify and champion organisational change

In this connotation, leadership is different from management:

Managers as administrators	Managers as leaders
…are heroes that have everything under control	…are able to act in ambiguity and uncertainty
…are the engines of the train and motivate as 'the best experts'	…look for imbalances, make them visible and address them
…lead others as superiors, they expect loyalty	…trust in self-responsibility, self-leading, self-organising, and believe that the system will find a way
…control the knowledge and allow overview only in the top hierarchy	…assure that the collected knowledge and the overview are IN the system
… work IN the system ('drive the car')	…work AT the system ('improve the car')

It must, however, be acknowledged that it is quite helpful to have sound management skills in order to become a leader. There are seemingly trivial things like regular meetings with documented results and their consequences, clear documented tasks, agreed and communicated roles, setting agendas, working with calendars or working with internal e-mails. Such 'management homework' can make a significant difference in improving the performance of an organisation.

These reflections lead us to the conviction that we have as much a 'water problem' as we have a 'water management and leadership problem'. Capacity building, therefore, goes beyond addressing technical skills by training. It needs to address the 'cha-cha-cha dancing skills' and work with managers towards leadership. This cannot be achieved by personnel development alone because leadership always develops in interaction with others and in the working environment ('in the action'). It must be accompanied by organisational development, action learning, reflective practice and transferring the learning into the organisation.

A Journey into Practical Experience

In this section, we report two exemplary (and true) stories of organisations in the water sector which show that there is room for improvement beyond technical solutions: in the field of capacity building and with – in these cases – a focus on leadership.

The Case of a National Water Agency

We are in a country which is hot and dry and water resources are threatened in terms of quantity and quality. The water sector has been highly centralised, run in an offer-oriented and state-driven fashion, focusing on building dams and other infrastructure works. Logically, the water organisations are found in the Ministry of Infrastructure and Equipment. The whole sector is run by engineers and technicians who had all rights to be proud of their professionalism. Regional organisations implement entities of central decisions.

Changes in politics and in the paradigms of the sector move a large part of the involved organisations under the responsibility of the Ministry of Environment. There, they are put under the authority of two different state secretaries. Simultaneously, the decentralisation process in the country creates regional semi-public water agencies. The roles of the central organisations change towards 'regulating, ensuring, coordinating, monitoring,...' The central level professionals feel disappointed in their professional conviction. They lack orientation, as the legal and regulatory framework is not finalised. Between the two state secretaries they lack a feeling of belonging. Therefore, they perceive themselves as victims of forces beyond their control. Leadership is preoccupied with survival under the new conditions. Fragmented and uncoordinated efforts to find and fulfil the new roles are the symptoms of these circumstances. The organisation is in a state crisis, is uncertain about its mission, compromises its values and does not have a clear vision. There is a general spirit of organisational demotivation paired with professional enthusiasm that cannot find its way. However, there is a solution in sight: 'We are sure that with the next elections this anomaly will disappear'.

The Case of a Regional Water Agency

A few months later, we visit a regional water agency. We are astonished to find attitudes and behavioural patterns similar to those of the central level. They see themselves as victims of higher forces beyond their influence: 'We should be autonomous, but the central level dictates our budget. Even though we are a semi-public agency by now, we continue to be paid like civil servants. Almost 50% of our staff has gone into early retirement and has not been replaced. So all we do is run after urgencies when they emerge'. When processes get stuck, the blame is generally put on external partners: 'We always wait for their reaction for many weeks'.

Of course, all personnel are engineers and technicians. The head of Department of Finances and Human Resources (an engineer with an extra training in accounting procedures) has been appointed the director. The department heads do not meet. When there are any meetings in the agency, people do not come ('because the provincial governor had called me'), and those who come are not sure why they are there: 'In any event it is a waste of time, as there are urgent things on my desk and nothing will come out of this meeting'. This is not surprising as there are generally no agendas, meetings are arranged at short or no notice, and there is no time management, no objective, no final reflection, no moderation, no documentation and, for sure, no impacts.

Furthermore at the regional level, solutions are at hand:

- 'Clearly, we need to recruit personnel, if only the budget would allow for that'.
- 'The central authorities have to...'.
- Big studies are given to consulting bureaus (however, the competences seem to get lost and it is not clear how such an external partner can be steered).
- There is a donor with a project that deals with a fragment of the agency's task (so they can stop thinking about it).
- There is another donor who comes with technical advice. They want to bring order into the fragmented data, create a functional web site (the existing one never worked) and install a document management system. However, the counterpart of the international expert never attends any meetings. The director does not react to this situation. Other staff, notoriously overworked, are shouldering the added burden.

In both cases, surprisingly the organisations experienced serious turbulence due to change and had to continue business as usual. There was little attention given to the process of change itself and hence no change in management. Reactions such as 'business as usual', 'diving into operational emergencies' or 'it will come back to normal' are typical in change processes, which, if not dealt with, effectively hinder the changes to be implemented.

The Need to Combine Personnel, Organisational and System Development

Based on our experience in capacity building and referring to the two examples cited in the previous section, we conclude that there is and there will continue to be a need to intervene with capacity building at three different and complementary levels:

- Individual level
- Organisational level
- System level

At the *individual level*, personnel development measures can target decision-makers, junior-level managers, senior and also junior professionals as well as support staff (a functional secretariat is a key factor in enhancing a decision-maker's efficiency and effectiveness). The organisational unit responsible for human resources development would set up a concept of personnel development by establishing a plan of personnel needs as well as a concept of requisitions to define concrete measures for recruitment, training and outplacement. In setting up such a concept, the HR unit would balance the individual perspective (abilities and inclinations) with the organisational perspective (internal and external challenges).

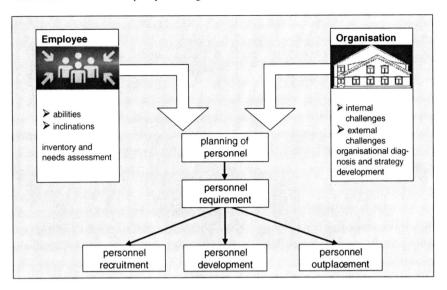

The actors in this process also include the top executives, the executive staff and the target groups. It can also involve higher levels of hierarchy and internal or external training providers. Concrete measures, finally, can involve a variety of training methods such as on/off/near-the-job training, individual and group training, job enrichment, enlargement and rotation as well as work in groups and projects (on-the-job). Other methods are mentoring and coaching.

In the two examples cited above, leadership training will be offered at the national and regional level. At the regional level this will be complemented by training in facilitation skills (to make meetings successful), in human resource development (to enable the HR unit to develop further concepts and measures on its own) and in a number of technical skills. By grouping training participants with participants from other regional agencies, the training cycle will support the creation of horizontal networks and thus contribute to system development in the institutional landscape.

At the *organisational level*, tailor-made development processes need to be identified. Therefore, we describe in a few words an exemplary process in the regional example cited above. The intervention (by external organisation development consultants) started with a rapid organisational appraisal and individual interviews at the executive level. This was followed by mini-training on management terms and models to clarify the existing imprecise use of management language. In the next step the executive staff and the director went through a self-assessment with regard to leadership (as part of the EFQM process). This created awareness of gaps in the present leadership process. The executive group met several times to clarify the mission and values of the agency. In the process, it became clear to the executives that the assumption of having a common understanding of mission and values had been erroneous. The discussions contributed to building identity and the spirit of a family of professionals. Simultaneously, a working group was trained in optimising processes. In both groups, the facilitation of the meetings shifted more and more to members of the organisation who were mentored in the process. Through these and similar interventions the director and the leader of the process group started to embrace their leadership role and found themselves in the driving seat. Thus, the ownership for the whole capacity building process was continuously given into the hands of the organisation. In order to create vertical links, to build a common language and to contribute to system building, it was decided to repeat the process at the national level and to exchange experiences with neighbouring regional agencies. The organisation development process is still ongoing.

At the *system level*, the institutional set-up of the sector needs to be developed. This can involve the legal and regulatory framework, restructuring, decentralisation, privatisation within the limits of ethics, national dialogue, the establishment of platforms of communication and so on. In the two examples cited above, studies are being commissioned with the aim of rearranging the sector. Through the organisation development work, the feeling of victimisation has been reduced in the two organisations. There is now a growing recognition of the opportunities to act within the boundaries of the system.

The InWEnt Capacity Building Concept

The concept of capacity building constitutes InWEnt's guiding principle in 'Strategic Human Resource Development' (SHRD). It defines measures in human resource development and in international cooperation. 'Capacity Building' encompasses advanced professional training, personnel and organisational development so as to achieve a given set of objectives. Capacity building also focuses on strengthening one's partners' capacity to plan and finally implement lasting development strategies and policies. In this sense, advanced professional training and personnel development as elements of SHRD finally determine the development of organisations through constant interaction. InWEnt systematically creates accessibility to knowledge available on a global scale, thereby rendering it utilisable to its partners as a strategic resource for development. InWEnt considers itself an international knowledge broker. By fostering dialogue, providing advanced training, building networks and providing advisory services for professionals and managers, InWEnt supports change processes in organisations and institutions in addition to reform processes in economy, politics and society worldwide.

InWEnt's work aims at fostering skills in problem-solving as well as applying acquired knowledge of decision-makers and executive personnel at all levels, as they perform change and reform processes. In addition, it intends to strengthen the structures required for these processes. InWEnt promotes the perception of societal responsibility and systematically creates accessibility to knowledge as a strategic development resource.

InWEnt understands societal change as a complex process. During this process, people in different capacities and different organisations try to determine and shape their working and living conditions efficiently and effectively by themselves. They effect these changes in a political realm by creating the conditions for economic, social, ecological and cultural development. Equally, they are also involved in organisations of civil society, business and municipal affairs. Participation that promotes change is, therefore, a prerequisite and a goal at the same time – for the political focus and for the methodical approach of InWEnt's SHRD.

InWEnt undertakes measures in the area of continuous education of personnel and organisational development. It recognises the need for systemic change management in its programme on three inter-connected levels: the individual, the organisation and the system.

At the *individual level*, InWEnt strengthens the competence and ability of decision-makers to act. This applies likewise to junior-level managers and senior professionals. This includes specific qualifications such as negotiation and communication skills aimed at interdisciplinary cooperation as well as competent management of knowledge and organisations. Of no less importance are qualifications in interdisciplinary thinking, having global perspectives and diversity management. As a result, InWEnt's programmes do not exclusively focus on conveying (relaying) knowledge but also on changing behaviour.

The methods focus on the target, application and the action. They promote the participation and self-learning competence of the participants and the exchange of experiences. The advanced training is practice oriented and includes learning projects, planning games, traineeships and internships, while e-learning is used systematically in the form of blended learning. Joint learning enables participants to experience intercultural dialogue first hand, and it is emphasised as a competence required in all societies. All participants are encouraged to help shape change processes, to improve their own action competence, while also expanding their social and intercultural competences and critical self-assessment.

Working with partner organisations, InWEnt develops strategies for advisory services for personnel development and *organisational development*. InWEnt supports the implementation of these strategies and helps to assess their impact. National and regional training institutes constitute InWEnt's crucial partners whose competencies are strengthened through 'Train the Trainer' programmes.

At the *system level*, InWEnt fosters the action and decision-making competence, as well as the perception of individuals in a position of responsibility. It achieves this objective by communicating with different political players and assessing the scope of action and viable alternatives, while also discussing strategies for policy shaping in terms of reaching international development objectives (such as the Millennium Development Goals). In doing so, InWEnt provides a valuable contribution to the international competence of decision-makers and to the strengthening of the partner country's institutional environment. In the realm of development cooperation, InWEnt aims to strengthen the development potential in our partner countries.

In its programmes, InWEnt uses different instruments for personnel and organisational development. In accordance with the joint goals set forth with its partners, InWEnt selects these instruments and combines and supplements them with one another.

These applied instruments by InWEnt include:

- Advanced education and training
- Dialogue
- Networking
- Advisory services for HRD

One Example: InWEnt's MENA Water Programme (2005–2008)

The objective is to support reform processes in strengthening professionals to contribute actively to the improvement of water sector performance. The programme targets junior professionals, decision-makers and senior experts of InWEnt's partner organisations in the water sector of Algeria, Egypt, Jordan, Morocco, Palestine, Tunisia, Syria and Yemen. All in all, InWEnt is working with around 65 national water organisations as well as the Arab Water Council on a transnational level.

Realising over 40 activities in 4 years, InWEnt is applying a systemic and flexible approach. Thus, the MENA water programme combines professional knowledge with methodical and regional competence and ensures an appropriate consideration of current needs and market trends.

Modern Capacity Building: Elements of the InWEnt approach

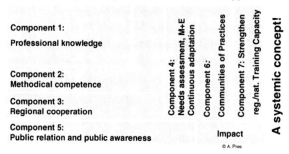

The MENA water programme consists of seven components, which are closely linked and complementary to each other, to continuously strengthen and broaden capacity:

- Component 1 – Professional knowledge: managerial knowledge upgrades in the context of water policy, integrated water resources management, water supply and sanitation and rural/agricultural water use

- Component 2 – Methodical competence: strengthening of governance competence by management of development and change processes, training of trainers and development of tool kits
- Component 3 – Regional cooperation: exchange platform for our partners
- Component 4 – Needs assessment, M+E: studies, assessments and adaptation
- Component 5 – Public relations and public awareness: information material and public awareness campaigns to inform and disseminate information on these issues
- Component 6 – Communities of practices: documentation of good practices, dissemination of lessons learned and exchange with capacity building players
- Component 7 – Strengthening training capacities: advisory services on organisational development for existing training entities

Component 1 focuses on professional knowledge transfer, emphasising applied knowledge and management issues within the water sector. Component 2 provides a sound methodical knowledge; it facilitates the application of knowledge and promotes the understanding of institutional complexity. Component 3 is the platform to exchange knowledge and to strengthen regional cooperation. Component 4 is the base for evaluation and modification of the capacity building offer, if necessary. Component 5 relies on the former components and disseminates knowledge for the public, which component 6 does for professionals. Component 7 was included after 18 months of implementation to increase the outreach of capacity building measures by strengthening training entities on site.

This systemic and flexible approach allows to react promptly on changes and to extend the offer of capacity building measures according to the needs of the partners involved as well as changing policy guidelines. Applying the mix of instruments and its sub-instruments (as dialogue workshops, seminars, short-term training courses, etc.), different levels of target groups can be reached and interlinked.

Conclusion

In the current development and globalisation processes that are engulfing humanity, change is the only stable pattern and this will only increase with time. People are at the centre of change. Leadership and management are essential in guiding these change processes successfully and in a sustainable manner.

In that perspective, SHRD has to focus on leadership and management training. Professional competence in the sense of technical skills is no longer enough. Promoting international learning processes shifts in this context more and more to strengthen competence to act effectively in changing environments.

SHRD is the core factor in achieving capacity building in global partnership focusing on the people actively engaged in shaping sustainable development. The attention focuses on executives and senior staff members as well as young professionals of relevant organisations, the organisations being those that promote reform processes. To achieve impacts, SHRD has to stress the links between the competence of individuals, the performance of organisations, and the relevant reforms of the political systems. Accordingly, SHRD becomes an essential instrument of capacity building measured on 'individual competence level' as output, on 'organisational performance level' as outcome and on development level in terms of impact.

In this context of capacity building through SHRD, creating local ownership in capacity development processes is essential. Without local ownership there is no effectiveness, no efficiency, no broad-based impact or significance and no sustainability.

SHRD is made up of analysing, advising and enhancing individual competencies by specific training which is necessary to change organisations, institutions and society. In this context training offerings have to be customised for clients' specifications and practical applications.

Apart from praxis-oriented training, the exchange of experience by networking and dialogue – on all levels and including public and private sectors as well as civil society – is most important not only for agenda setting and development of roadmaps but also to *widen* systematically the *access to knowledge and innovation*.

The worldwide collaboration of training institutions aiming at SHRD under a strategic alliance should be a further important approach to achieve human capacities' objectives more efficiently at present and in future, by 'strengthening capacities to build capacities' on national and regional levels.

Bibliography

InWEnt-Internationale Weiterbildung und Entwicklung (2006a) Capacity building concept. Bonn
InWEnt-Internationale Weiterbildung und Entwicklung (2006b) Didaktik Leitfaden. Bonn

The European Water Framework Directive: Potential for Change and Implications Beyond 2020

José Albiac and Juan Ramón Murua

Introduction

The problem of water stress and water quality is one of the main environmental issues in Europe, together with climate change, air quality, biodiversity and soil quality. With respect to water quantity, withdrawal of water resources in Europe is above 20% of renewable freshwater resources. The major pressures on water quantities occur during summer in southern countries because of irrigation demand and tourism activities. In the coming decades, the likely increase of withdrawals and climatic change will result in more intensive pressures on water quantities in these southern countries.

With respect to water quality, the problems are driven by pollution of water resources. The pollutants are nutrients and organic matter and dangerous substances such as heavy metals and chemical compounds. The nitrate emissions from agriculture have decreased somewhat in the last decade in most rivers, but there still remain problems of eutrophication and pollution of drinkable water. There are fewer rivers strongly polluted as a consequence of reduction in organic matter loads, use of detergents free of phosphates, and operation of new treatment facilities in urban centres. However, around 20% of European surface waters still have severe pollution problems.

The European Framework Directive, approved in 2000, is an important initiative of the European Union, intended to protect all continental, groundwater and coastal waters. It has a great potential to solve water scarcity and water quality problems in Europe by 2021, when the first management cycle deadline ends, and by 2027 which is the final deadline for meeting the objectives of the Directive.

However, there are two aspects in water resources management that are difficult to solve and may hinder the attainment of the objectives. One is the sustainable management of aquifers, and the other is nonpoint pollution control, and the cause of the difficulties is that they are common pool, resources where implementation of

---

J. Albiac (✉)
Department of Agricultural Economics, Agrifood Research and Technology Centre (CITA-DGA), Av. Montañana 93c, 50059 Zaragoza, Spain
e-mail: maella@unizar.es

policy measures is not an easy task. The design of reasonable measures for the management of aquifers and nonpoint pollution requires information and knowledge on biophysical processes linked to aquifer dynamics and pollution transportation and destination. Generating this information and knowledge is demanding in resources and takes time. The design of measures by the water authorities must also take into account the strategic behaviour of stakeholders. Aquifer management and nonpoint pollution control involve cooperation among stakeholders (Albiac et al. 2008), and this is a daunting task for water authorities. Managing common pool resource has been a failure worldwide.

Water Demand by Sectors and Water Scarcity Problems

Water abstractions in Europe for all uses reach at present 307,000 hm^3, of which 115,000 hm^3 are used in agriculture, another 104,000 hm^3 for cooling and electricity production, 53,000 hm^3 for urban demand, and 35,000 hm^3 for industrial demand (Table 1). Water for cooling and electricity production returns to water courses with small changes in quality. However, most water is used for agricultural, urban, and industrial purposes, which degrade the quality of water returns. These consumptive uses generate water stress in many European regions and problems of point and nonpoint pollution of water courses. There has been a reduction of total water abstractions in the last decade, although tendencies differ widely among sectors. The likely outlooks for next decades are increases in water use by agriculture and industry, strong reductions for cooling and electricity production, and stability in urban water use (EEA, 2005a).

Table 1 Water use in selected European countries (2001)

Country	Total water extractions (hm^3)	Urban use (hm^3)	Industrial use (hm^3)	Irrigation (hm^3)
Bulgaria	5,800	1,100	300	900
France	33,500	5,800	3,600	4,800
Germany	40,400	5,500	5,600	600
Greece	8,900	900	100	7,700
Hungary	5,600	700	200	500
Italy	56,200	10,100	9,600	25,900
Poland	11,600	2,200	600	1,000
Portugal	9,900	800	400	8,800
Romania	7,300	2,500	900	1,000
Spain	37,700	3,800	1,400	24,600
Turkey	39,800	4,300	3,500	31,000
United Kingdom	15,900	6,300	1,600	1,900
Total Europe	307,200	53,300	34,900	115,100

Sources: EEA (2005b); INE (2004; 2005); IFEN (2005)

Agriculture accounts for more than a third of total water extractions, and the volume of irrigation water will grow to cover the expansion of irrigated acreage in the south of Europe, Hungary, and EU candidate countries such as Turkey. Economic development may increase the use of water by the industrial sector, in particular in eastern countries and EU candidates. The use of water for cooling and electricity production will be cut by half as a result of more efficient refrigeration systems in power-generating facilities. The new tower cooling systems reduce the amount of water by two orders of magnitude per megawatt-hour, compared to current refrigeration systems with single circulation. Urban demand represents 20% of abstractions, and its evolution will be stable since it depends on countervailing factors such as household size and income, water prices, and technological changes that improves water efficiency.

In central and northern European countries such as Germany, France, and UK, the main water abstractions are for power generation, which will fall strongly in the coming decades, while industrial use may increase. In contrast, the main water use in southern countries is irrigation, with joint abstractions by Spain, Italy, and Turkey above 80,000 hm^3 (Table 1). Irrigation water demand would increase as a consequence of the expansion in irrigated acreage and also because of the impact of climatic change on irrigation water crop requirements in the coming decades. Urban and industrial water demand will increase in eastern countries and Turkey, following the rise in household incomes and industrial activities up to western countries' levels.

In summary, northern and central European countries do not face problems of severe water stress, and their main extractions are used for power generation which return to watersheds. These extractions are going to decrease substantially and the outlook for the next decades is less water stress in these regions. This is the case for the rivers Rhine, Elbe, Loire, Vistula, Oder, Rhone, and Garonne. The more serious problems of water scarcity take place in the arid and semiarid regions of southern Europe, such as the southern half of the Iberian and Italian peninsulas and the Anatolian peninsula. The use of irrigation water is very high in these regions, and scarcity problems will worsen because of the expansion of irrigated acreage and the increase of water demand for tourism activities in coastal zones. In the coming decades, the effects of climate change would also have a negative impact on available water resources in Mediterranean countries.

Water Quality Problems

Surface, groundwater and coastal waters have different uses, including domestic, industrial, irrigation, recreation and support of aquatic ecosystems. Human activities are linked to water and land resources and generate wealth, but these activities also degrade water quality through point and nonpoint pollution. To cope with this water degradation, different quality standards have been implemented depending on the final use of water. There are two alternatives to reach the appropriate quality standard: one is to reduce the pollution loads at the water courses and the other is water treatment of the waters being used. The more demanding standards are those for drinking water.

The volume of wastewater increased considerably during the last century due to industrial development and the growing consumption of households. The effects of discharge of wastewaters depend on the sewage network and treatment facilities, the industrial production processes, and the type of products consumed by households. In recent decades, there has been a surge in the urban population linked to sewage networks and treatment facilities, although there are considerable differences among European regions. Almost all the population in northern European countries is connected to water treatment facilities, but only half the population is connected in the new member countries of the European Union.

The Urban Wastewater Treatment Directive, passed in 1991 and modified in 1998, required building treatment plants with secondary treatment in urban centres with population higher than 15,000 inhabitants by 2000 and higher than 2,000 inhabitants by 2005. The central and northern European countries have already treatment plants with secondary and tertiary treatment. Tertiary treatment is more advanced than secondary treatment and reduces the emission loads of the nutrients phosphorus (up to 60%) and nitrogen (up to 90%). Tertiary emission loads of phosphorus and nitrogen are 0.1 kg and 2 kg per person per year, respectively. Countries in the south of Europe, together with France, Belgium, and UK, only have treatment plants with secondary treatment, and the emission loads of phosphorus and nitrogen are 0.4 kg and 3 kg per person per year, respectively (EEA, 2005a).

The Urban Wastewater Treatment Directive has achieved a significant reduction of polluting emissions on surface waters, curbing the environmental damages on aquatic ecosystems. However, the level of emissions from treatment plants remains high and may cause eutrophication problems in vulnerable areas.

The number of dangerous substances that may affect water quality is high, with very different sources. The manufacturing industry is responsible for most of the emissions of heavy metals (lead, mercury, cadmium), while other substances such as nutrients and pesticides come basically from agriculture. A few substances have been regulated in the last decades resulting in a reduction of their emissions, but the emissions abatement is not general. Table 2 shows pollutant concentrations in selected European rivers. There are important pollution loads by nutrients (nitrates and phosphorus) in rivers Thames, Guadalquivir, Seine, and Escaut, and a high concentration of heavy metals in rivers Seine, Tajo, Guadalquivir, and Porsuk.

There has been a reduction of phosphates in detergents used by households in the last few years, with a fall in the phosphorus load in treatment facilities from 1.5 to 1 kg per person per year. Meanwhile, the nitrogen load has remained constant at 5 kg per person per year. The phosphorus loads received by water courses originate from urban and industrial point sources and agricultural and livestock nonpoint sources, while most of the nitrogen loads come from nonpoint agricultural and livestock sources.

Although information on the status of aquatic ecosystems in Europe is quite scarce, it seems that the water quality in some rivers is improving. The improve-

Table 2 Water quality in selected European rivers (average 2002–2004)

Country	Watershed	BOD (mg O$_2$/L)	Nitrates (mg N/L)	Phosphorus (mg P/L)	Lead (µg/L)	Cadmium (µg/L)	Chromium (µg/L)	Copper (µg/L)
Norway	Skienselva	2.0[a]	0.2	0.01	0.2	0.02	0.11	0.62
Sweden	Dalalven		0.1	0.02	0.5[a]	0.02	0.37[a]	1.48
Denmark	Gudena	1.9	1.3	0.09				
UK	Thames	3.4	6.6	0.66	2.9	0.10	1.17	6.63[a]
The Netherlands	Maas	2.5	3.6	0.21	2.8	0.15	1.77	3.77
Belgium	Escaut	3.6	4.7	0.66	12.0	0.67	9.93	10.10
Germany	Rhein	3.0	2.5	0.14	3.0	0.20	2.55	6.22
	Elbe	6.9	3.0	0.17	2.2	0.18	1.20	4.36
	Weser	2.8	3.7	0.14	4.5[a]	0.20	2.03[a]	3.56
France	Loire	3.2	3.1	0.21		0.40[a]		
	Seine	3.1[a]	5.6	0.63[a]	22.1[a]	2.18[a]	24.67[a]	15.03[a]
Spain	Guadalquivir	4.2[a]	6.1[a]	0.95[a]	10.2[a]	1.87[a]		5.73[a]
	Ebro	1.9	2.2	0.09	7.5	0.23[a]	0.92[a]	1.61[a]
	Guadiana	1.6	1.8	0.69[a]		3.39		
Portugal	Tejo	2.3	1.0	0.20	11.0	3.00	22.33[a]	2.10
Italy	Po	1.3	2.5	0.25				
Greece	Strimonas		1.8	0.14		0.64[a]		
Turkey	Porsuk	1.4	1.5	0.06	12.2	6.50	7.50	5.67

Source: OECD (2007).
[a] The average is for years 1999–2001 or before. The Biochemical Oxygen Demand (BOD) measures pollution by organic matter, and water is considered drinkable for BOD of 0.75–1.50 mg O$_2$/L

ment results from the abatement of emissions of organic matter and phosphorus linked to new treatment facilities in urban centres and the abatement of heavy metals and chemical substances undertaken by industries. However, the nitrogen and phosphorus loads coming from agricultural nonpoint pollution are not controlled, and the relative importance of this pollution is increasing. Thus, between 50% and 90% of the nitrogen loads in surface waters come from agriculture. Pollution problems from agricultural sources are characterised by the uncertainty of the source location and by the impossibility (or very high cost) of measuring the emission loads of each farmer. This question has important implications for the design of pollution abatement measures, since point pollution control measures are not useful, and more sophisticated measures are required.

The intensive use of fertilisers is a severe problem in central and northern European countries. Fertiliser consumption in these countries is above 150 kg/ha, while consumption in southern countries is below 150 kg/ha. Fertiliser consumption corresponds to the sum of nitrogen (N), phosphorus (P_2O_5), and potassium (K_2O). Fertiliser consumption is above 200 kg/ha in Germany, Belgium, France, the Netherlands, Ireland, and UK. For example, the nitrogen surplus in soils is 215 kg/ha in the Netherlands and 100 kg/ha in Belgium and Germany, compared to 40 kg/ha in Spain (EEA, 2003), and this surplus is the origin of the nitrate pollution in water. Therefore, the problems of water quality from agricultural nonpoint pollution are more serious in central and northern European countries, while the main problem in southern countries is water scarcity.

Concern on water scarcity and quality has resulted in the development of an extensive body of rules and regulations in the European Union: the Water Framework Directive (2000) and the directives of Drinking Water (1998), Integrated Pollution Prevention and Control (1996), Urban Wastewater Treatment (1991), Nitrates (1991), Dangerous Substances (1976, integrated in the Water Framework Directive in 2006), and Bathing Water Quality (2006).

This legislation has attained important results in curbing point pollution from urban and industrial sources, due to the construction of treatment facilities in urban and industrial centres and the fall in the emissions of dangerous substances from industrial processes. The consequence has been an improvement of the quality of surface and coastal waters and less pressure on aquatic ecosystems. However, the problems of agricultural nonpoint pollution remain, in particular those of nutrients and pesticides (European Commission, 2002), and also the problems of water scarcity in Mediterranean countries.

The Water Framework Directive

The European Union approved an important legislation to protect water resources, the Water Framework Directive, which was subsequently enacted in European Union member countries. The Directive creates a common framework of action in water policy, with the objective of protecting continental surface waters, transitional waters, coastal waters, and groundwaters. This protection intends to avoid any further degradation of water quality and to improve the aquatic ecosystems conditions, promote the sustainable use of water resources, protect and improve water quality through the abatement of emissions and discharges, reduce gradually the pollution of groundwaters, and finally contribute to curtail the effects of floods and droughts. Water management is organised at the level of river basin district. The Directive aims at securing a sufficient supply of surface and groundwater in good condition, attaining a balanced and equitable supply, and contributes to a significant water pollution abatement.

European countries have defined their river basins and basin authorities by 2003 and have completed the characterisation of pressures, impacts, and economic analysis of basins by 2004. The results have been used to evaluate the impact of human activities and to identify the areas requiring special protection, guiding the elaboration of the basin management plans and the programmes of measures by 2009. Water pricing policies will be introduced in 2010, and the programmes will be operational in 2012, in order to reach the environmental objectives in 2015.

The Directive introduces the principle that water pricing should be close to full recovery costs to improve the efficiency in the use of water. The full recovery cost must include the abstraction, distribution, and treatment costs and also the environmental costs and the resource value. The Directive establishes a combination of emission limits and water quality standards, with deadlines to achieve appropriate quality for all waters ("good ecological status"). Water management should be based at the river basin level and consider stakeholder participation. Water prices paid by users should approach full recovery costs.

The principle of cost recovery is one of the key elements in the economic analysis advocated by the Directive. The increase in water prices up to recovery costs is a very interesting measure in the industrial and urban sectors, since the industrial and urban water demands respond to water prices, hoping to achieve higher efficiency in water use. In contrast, water demand in irrigation does not respond to water pricing, and this questions full recovery costs in irrigated agriculture as a valid alternative for water quantity assignment.

Setting some minimum price levels for water for irrigation will make farmers understand that water is not a free good. However, using water pricing as a mechanism to allocate water in irrigation is questionable. Cornish and Perry (2003) and Bosworth et al. (2002) show compelling results from the literature and from empirical studies that demonstrate the impossibility of using water prices to assign water in irrigation, both in developed and developing countries. As alternative to water pricing, these authors indicate that introducing water markets is much more reasonable, although difficult to implement. Therefore, the emphasis of the Directive on water prices has no impact to reduce irrigation demand in Mediterranean countries, and Spain is a clear example of this as discussed in the next section.

In order to reach the objectives of the Water Framework Directive, the best alternative to reduce the water scarcity caused by urban and industrial demands is water pricing. Collective irrigation systems based on dams and canal networks should be controlled through command and control measures, while irrigation districts based on individual pumping from aquifers need sophisticated incentive schemes that promote the cooperation of farmers in water conservation.

Additional measures against scarcity include reuse of water from treatment plants and seawater desalination, although their use is quite limited at present. Another option is improving the distribution networks, since their condition affects largely the total water extractions required to cover the demand of the different sectors. Water losses in channeling networks are substantial in many European countries, and upgrading the facilities would involve large savings but also large investments.

There are some important methodological and information-related problems within the policy analysis of the Water Framework Directive, since many basic concepts of environmental policy analysis are not well understood.[1] The emphasis of the Water Framework Directive on water pricing in order to achieve water use efficiency and protect the resource followed the Dublin Declaration of 1992, but it is a flawed approach. The problem with this "economic good" approach assumed by the WFD and by many environmental consultants and decision makers in Europe is that the price mechanism can work only where water is a private good (rivalry in consumption and exclusion) which is traded in markets.

Domestic and industrial uses have characteristics of a private good, but irrigation is different because it can be considered as an impure public good, with environmental externalities. Water pricing could modify consumption where markets exist, as in urban networks for domestic and industrial demands, but not in agricultural or environmental uses. Furthermore, water markets cannot internalise environmental externalities, as the California and Australia cases show. Protection and conservation of water resources, which are common pool resources, require the cooperation of the several parties managing the resource to achieve collective action.

Another difficulty is the lack of basic statistical information and knowledge on biophysical processes, which favors the strategic behaviour by countries, basins, and stakeholders in the entire implementation process of the Directive.

The basic and supplementary measures in the Water Framework Directive are not applicable[2] because they do not take into account the state of knowledge in policy analysis from the field of environmental economics. The Directive does not consider the concepts of private good, public good or externalities, and therefore ignores that different types of measures are needed for different types of problems in water resources. The conceptual and empirical misunderstanding in the policy analysis of the Directive is such that there is a large confusion among the key water consultants and environmental decision makers.

---

[1] Such as objectives, instruments (institutional, economic, command, and control), welfare optimum, target, cost-efficiency, private good, common pool resource, stakeholders cooperation, and collective action.

[2] The definition of "basic measures" in the WFD shows that they are not policy measures at all, but a reformulation of the objectives that are supposed to be reached with previous water legislation. The definition of "supplementary measures" in the WFD is overly general and does not have any practical application.

To improve social welfare and policy measures, knowledge is needed on biophysical processes, on social benefits and cost functions for each alternative, and on the optimum social welfare derived from these functions. Without knowing the benefit and cost functions for each alternative, decision makers can develop policies by considering cost-efficiency. Nevertheless, the WATECO Committee in charge of the WFD economic analysis has decided to ignore all these issues and consider as environmental costs whatever environmental expenses countries decide to present.

In order to elaborate reasonable alternatives, it is essential to clarify the conceptual methodology of policy analysis and determine the requirements regarding water statistics and scientific knowledge on biophysical processes that are needed. However, even when biophysical knowledge is generated, the management of water resources is still quite a challenge, because of the public good and environmental externalities aspects of water. The incentives from policy measures should address the strategic behaviour of stakeholders, in order to encourage cooperation and collective action in conserving water resources.

Another issue is that water institutions are very weak or nonexistent in most European countries, and these countries do not have experience in collecting data and designing and implementing reasonable water policies that work. Examples are Germany and Italy, which do not have water authorities at basin and federal levels. In the case of Germany, each state has its own methods, databases, and assessment approaches, and achieving policy coordination remains to be seen. In the case of Italy, there is a plethora of very small organisations in charge of water, which lead to very serious information shortcomings on water resources, and also to difficulties on policy design and implementation.

Applying the Water Framework Directive: The Case of Spain

Water resources extraction and utilisation by sector in Spain are presented in Table 3. Extractions are close to $40,000\,\text{hm}^3$, of which $6,200\,\text{hm}^3$ are used for cooling in electricity production, and $32,000\,\text{hm}^3$ cover the demand for irrigation, water supply companies, and other industrial and service sectors. Losses in primary and secondary distribution networks are high and reach $5,500\,\text{hm}^3$. Household demand is $2,600\,\text{hm}^3$ with an average price of $1\,\text{euro}/\text{m}^3$, and industrial and service demands are $3,200\,\text{hm}^3$ with an average price of $0.25\,\text{euro}/\text{m}^3$. Net irrigation demand is $20,700\,\text{hm}^3$, and prices are related to the type of agriculture. In inland irrigation areas with collective systems of dams and canals, and field crops of low profitability, prices are close to $0.06\,\text{euro}/\text{m}^3$. In the irrigation areas of eastern and southeastern Spain with individual pumping from aquifers and high-profit crops, the range of prices is between 0.09 and $0.21\,\text{euro}/\text{m}^3$.

The growing pressure of these economic activities has created problems of water scarcity and quality degradation, mostly linked to groundwater. The most severe problems are located in southeastern Spain, with pressures coming from agriculture, urban growth, and tourism on the Mediterranean coast. In inland Spain, surface

Table 3 Water resources extraction and utilisation by sector in 2002 (hm^3)

	Total	Agriculture	Water companies	Other sectors	Cooling
Extractions	38,200	25,200	5,400	1,400	6,200
Surface	32,500	20,900	4,200	1,200	6,200
Groundwater	5,700	4,300	1,200	200	
Network losses	5,500	4,500	1,000		
Utilisation					
Agriculture	20,700	20,700			
Households	2,600		2,600		
Other sectors	3,200		1,800	1,400	
Cooling	6,200				6,200

Sources: INE (2005) and Martínez and Hernández (2003). Figures do not include hydropower extractions, estimated by MIMAM (2000) at an average of 16,000 hm^3

water resources are under the effective control of basin authorities that manage resources wisely.

The European Water Framework Directive approved in 2000 was enacted in the Spanish legislation in 2003, just after approval of the Spanish National Hydrological Plan (2001) and National Irrigation Plan (2002). The National Hydrological Plan involved large investments (19 billion euros) aimed at increasing water supply for agricultural, urban and industrial uses. Its main project was the Ebro interbasin transfer from northeastern to southeastern Spain, to alleviate the severe degradation of water resources in the area. The National Irrigation Plan involves investments (6 billion euros including the recent Irrigation Plan developed on an urgent basis) to modernise the largely outdated irrigation facilities, in order to save resources, enhance competitiveness and reduce pollution.

The National Hydrological Plan was subsequently modified in 2005, substituting the large Ebro water transfer that was its main project by the AGUA project based on expanding water supply through seawater desalination. Both versions of the National Hydrological Plan, with the Ebro transfer or with the AGUA project, maintain the traditional approach of increasing water supply.

The National Irrigation Plan has a good potential of saving water and curbing pollution through investments in advanced irrigation technologies. These investments do not guarantee the solution to all problems, but it is obvious that technical innovations in irrigation systems facilitate the private and public control of water quantity and quality. Achieving the National Irrigation Plan will require strong coordination between water authorities and irrigation water user associations.

One reason for coordination is the issue of water returns, in addition to the invesment in networks and plot irrigation systems. Water losses in distribution canals and plot irrigation systems return to watersheds, and when water losses are reduced after upgrading networks and irrigation systems, the problem that may appear is that farmers use the saved water in more water-demanding crops or in expanding irrigation land. The consequence could be an increase in evapotranspiration and reduction of water flows into the watersheds. A better solution is reducing water concessions to countervail the eventual evapotranspiration increases.

Water resources degradation in southeastern Spain is driven by intensive agriculture based on individual abstractions from aquifers, urban development, and tourism over the Mediterranean coast. Aquifer overdraft reaches 700 hm^3, in the Júcar (160 hm^3), Segura (220 hm^3), Sur (70 hm^3) and upper Guadiana (220 hm^3) basins. The massive overdraft is the consequence of decades of groundwater mismanagement, despite the fact that groundwater was declared public domain in 1985. Registration of both concessions and private rights of groundwater is far from complete, and the number of illegal wells could be above one million. In contrast, water scarcity and degradation is rather moderate in inland Spain because irrigation is based on collective systems: basin authorities control concessions, river flows, and dam reserves, while irrigation user associations manage irrigation districts. The experience and efficiency of this institutional setting ensures ecological flows and the management of droughts and floods.

The basin authorities in southeastern Spain do not control the number of wells or the volume of individual extractions from aquifers associated to very profitable crops, and hence they cannot impose recovery costs. Furthermore, the required price level to curb demand in these areas is above 3 euros/m^3, which is politically unfeasible (Albiac et al. 2006). On the contrary, basin authorities may impose any water price in the areas of inland Spain based in low profitable crops, because they have absolute control in collective irrigation systems. But the question is then the following: why should they play around with allocating water through water pricing when they can make direct and wise water allocations?

Another constraint to implement water pricing in irrigation comes from the results of the studies by Martínez and Albiac (2004, 2006), showing that water pricing is the less cost-efficient measure to abate nitrate pollution from agriculture.

There are some examples of unconvincing water policies being applied in Spain. One is the Plan of the Upper Guadiana, recently approved. The plan aims at curbing overdraft in the Western La-Mancha aquifer and recovering the Tablas de Daimiel natural park, one of the main wetlands in the country. Previous efforts to control illegal abstractions were turned down by the central Spanish administration, which sent the wrong signal not only to those exploiting illegal wells, but also to those with legal wells pumping in excess and depleting the aquifers. Instead of curtailing abstractions, the plan anticipates investments of 5 billion euros to eliminate 220 hm^3 of overdraft. What is surprising is that no economic valuation study has been undertaken on the environmental damages caused by the loss of this wetland, which could justify these large investments. Furthermore, the large investments in the Upper Guadiana will not work without careful designed incentives to gain farmers' cooperation. If the planned approach is generalised to the 500 hm^3 of aquifer overdraft in the Júcar, Segura, and Sur basins, then the additional investments needed would amount to 11 billion euros (Albiac et al. 2007).

A second example of a questionable water policy is the current AGUA project. The AGUA project includes investments of 1.2 billion euros to build desalination plants and increase supply by 600 hm^3, of which 300 hm^3 are for irrigation purposes in the coastal fringes. Although there is a potential irrigation demand in the area from greenhouses and other high-profit crops, the pumping costs are much lower than desalination costs, and farmers will not buy desalinated water. Public investments in desalination are only justified if basin authorities are able to enforce strictly a ban on aquifer overdraft, forcing farmers to buy desalinated water. However, the decision of the water authorities has been to subsidise desalinated water up to the level farmers are willing to pay (pumping costs).

An aspect of water management in Spain that should be stressed here is the institutional, technical, and organisational efficiency of basin authorities dating back 100 years. Basin authorities in Spain (Confederaciones Hidrográficas) have a richness of information which is lacking in most European countries, and they are very competent in managing surface water. There is also a high level of competence in the water business sector (construction, distribution, treatment, and desalination) and in the dynamic irrigation agriculture of southeastern Spain.

The problem to achieve sustainable water management in Spain is not lack of technical capacity, physical capital, or human resources, but the absence of political will in the design and implementation of reasonable policies. Solving the degradation and mismanagement of water resources in southeastern Spain is the key issue for moving toward sustainable management of water resources. Any supply-side policy of expanding water availability, such as the former Ebro interbasin transfer or the current AGUA project, is questionable as long as groundwater mismanagement continues. Demand-side policies such as forbidding aquifer overdraft or taxing water abstractions are technically and politically unfeasible, because basin authorities can only deal at present with surface water. Although there are informal water transactions in southeastern basins, the introduction of formal water markets requires enormous and persistent efforts. The Water Law was modified in 1999 to promote formal water markets, but it has not resulted in any significant transaction in almost 10 years. In any case, the introduction of formal water markets would require the control of groundwater. The experience of water markets in Australia and California demonstrates that economic instruments alone fail to protect water resources, and therefore command and control as well as institutional instruments have an important role to play.

The tasks ahead for basin authorities in Spain are quite challenging, since aquifers are common pool resources with impure public good characteristics (rival but non-excludable) and with environmental externalities. Their sustainable management requires that public authorities set up incentives that give rise to cooperation among agents managing the resource, in order to achieve the collective action needed for water conservation.

Conclusions

One of the important environmental questions in Europe is the scarcity and degradation of water resources. In Europe, the annual extraction of freshwater attains 20% of renewable resources, and the main pressures derive from the urban, industrial, and irrigation consumptive uses. These uses create water scarcity in some regions and a widespread water quality degradation from point and nonpoint pollution. Water scarcity is a serious problem in southern European countries, with a strong demand during summer for irrigation and tourism. Water quality degradation is driven by human activities which generate pollution from nutrients, organic matter, heavy metals, and other chemical by-products.

There are no serious problems of water scarcity in northern and central European countries, and their main extractions for energy production are going to diminish. In the semiarid regions of Mediterranean countries, such as the southern half of the Iberian, Italian, and Anatolian peninsulas, there is a massive use of water for irrigation. In these regions the scarcity outlook will dim because of expanded irrigated acreage and tourism in coastal areas and because climate change will reduce available resources.

The industrial development and the growing consumption by households during the last century explain the strong degradation of water resources. The efforts to curb pollution in western Europe were started in the 1970s through several European directives. This legislation addressed the effects of point pollution emissions from urban and industrial discharges, which depend on sewage collection and treatment facilities.

Despite these efforts undertaken by public administrations in the last decades, pollution by nutrients and heavy metals remains high in many watersheds of the more important river basins in Europe. The extensive European regulation has facilitated large investments in water treatment plants and technological innovations in industries and households, which have limited or reduced the emissions of some pollutants, but the abatement of emissions is not general. The efforts on urban and industrial point source emissions should continue, and effective control on nonpoint pollution is needed such as abatement of nutrients and pesticides from agriculture.

The future of water resources in Europe would depend on the management measures taken to solve the different problems in each European region. Water scarcity in southern Europe could worsen considerably by further uncontrolled extractions and the effects of climate change. Solving the scarcity problem may require reallocating some water from off-stream use by agricultural, urban and industrial uses to environmental uses both in aquifers and streams, and also in the coastal wetlands. There are serious problems of water quality degradation in almost all European countries, although their characteristics depend on the local pressures of human activities and the measures being taken in each region.

The case of Spain shows that the implementation of the Water Framework Directive is not an easy task. Both the Spanish Ministry of Environment and the European Commission Environment Directorate advocate water pricing in irrigation, using the Common Agricultural Policy to penalise farmers. Unfortunately, research projects funded by the European Commission and some other studies also recommend these flawed policy options.[3]

---

[3] An example is Downward and Taylor (2007) on Almería, which states that sustainable management can be achieved by water pricing and augmenting water supply through desalination. Irrigation water use in Almería is around $260\,hm^3$, and domestic and industrial uses are around

The problems of scarcity and quality degradation cannot be solved with these two policies. Water pricing is a very good instrument for industrial and domestic demands, but it is not applicable for irrigation. Water pricing is not an implementable option because: (1) there is no control on the huge number of illegal wells and the quantities pumped from aquifers; (2) water shadow prices are above 3 euros/m^3, a price politically unfeasible since desalination costs are 0.50 euro/m^3 and urban water prices are around 1 euro/m^3; and (3) the administration lacks the information on aquifer dynamics which precludes the enforcement of sustainable extractions.

The Common Agricultural Policy is also of no use to influence water extractions in southeastern Spain. CAP subsidies are targeted towards continental products such as field crops, while production in the area includes Mediterranean crops such as fruits and vegetables which have negligible CAP subsidies.

The design and implementation of reasonable measures required by the WFD is a difficult task not only in Spain or the Mediterranean member countries but also in the whole European Union. The improvement in the management of water resources involves better information and knowledge on surface and groundwater resources and on their associated ecosystems. These tasks need time and resources because of the complex biophysical, spatial, and intertemporal dimensions involved. At present, data on water quantity are not very good in the European Union, and data on water quality are even more limited. The quantity figures of the European Environment Agency do not match national figures (for example, Spain and France), and water quantity information from countries such as Italy is not available.

The policy analysis of the WFD needs substantial improvement in both the methodological approach and the choice of instruments. The "economic good" perspective that follows the Dublin Declaration is flawed, because the price mechanism can work only when water is private property. This may be the case in urban networks for domestic and industrial demand, but not for agriculture or environmental uses. Additionally, water markets cannot internalise environmental externalities. The common pool characteristics and environmental externalities of water resources call for cooperation and collective action by stakeholders and not for economic instruments.

The decision by the WATECO Committee (in charge of the WFD economic analysis) of taking as environmental costs, whatever environmental expenses countries have, highlights the weaknesses and drawbacks of the current water policy analysis in Europe. WATECO ignores both the principle of welfare optimisation derived from

---

90 hm^3. Water pricing may reduce industrial and domestic demand, but not irrigation aquifer pumping. Since the growing urbanisation pressure on the coast will take over any water pricing savings in industry and urban demands, scarcity from irrigation aquifer overdraft will continue. Desalination cannot work either, because farmers will not buy desalinated water unless a strict enforcement of overdraft is in place, a daunting task for authorities. The implication is that the measures advocated by Downward and Taylor cannot deliver the collective action required for water conservation. Examples from EU research projects advising questionable policies are WFD meets CAP (www.ecologic.de/modules.php?name=News&file=article&sid=1369), Aquamoney (www.aquamoney.org), AquaStress (www.aquastress.net), WADI (www.uco.es/investiga/grupos/wadi), POPA-CTDA (www.popa-ctda.net), and POLAGWAT (http://susproc.jrc.es/docs/waterdocs/FinalRep150802.pdf).

the benefit and cost functions of measures and also the principle of cost-efficiency used when the benefit function is unknown.

Knowledge of the underlying biophysical processes is critical for water management, specially for managing aquifers and controlling nonpoint pollution, and this requires the availability of basic facts on aquifer and pollution characteristics and dynamics at local watershed scale. Regarding pollution, information is needed on the emission loads, pollutants transportation processes, and the environmental pollution in watercourses. Also, the lack of economic valuation of damage costs to aquatic ecosystem from aquifer overdraft and nonpoint pollution precludes the assessment of the benefits of policy measures.

Even when all the biophysical knowledge is available, managing the quantity and quality of surface and groundwater is quite challenging because of the public good characteristics of water and the associated environmental externalities. The design of measures must take into account the strategic behaviour of water stakeholders, setting up incentives for cooperation in order to achieve water conservation through their collective action. Both aspects, biophysical knowledge and collective action, are unlikely to be in place by 2020, not only in Europe but worldwide.

Most European countries have no experience in collecting data to design and implement reasonable water policies, because their water institutions are weak or nonexistent. Two examples are Germany and Italy, which do not have water authorities at basin or central government level. The potential for change in European water policies points towards mild improvements by 2020. After 2020, the required institutional setting and collective action by stakeholders could be progressively achieved, but then climate change impacts would be a real challenge calling for a major improvement towards sustainable water management.

References

Albiac J, Martínez Y, Tapia J (2006) Water quantity and quality issues in Mediterranean Agriculture. In: OECD (ed) Water and Agriculture: Sustainability, Markets and Policies, OECD, Paris

Albiac J, Martínez Y, Xabadía A (2007) El desafío de la gestión de los recursos hídricos. Papeles de Economía Española 113:96–107

Albiac J, Dinar A, Sánchez-Soriano J (2008) Game theory: A useful approach for policy evaluation in natural resources and the environment. In: Dinar A, Albiac J, Sanchez-Soriano J (eds) Game theory and policy making in natural resources and the environment. Routledge explorations in environmental economics, Routledge, Abingdon

Bosworth B, Cornish G, Perry C, van Steenbergen F (2002) Water charging in irrigated agriculture. Lessons from the literature. Report OD 145, HR Wallingford, Wallingford

Cornish G, Perry C (2003) Water charging in irrigated agriculture. Lessons from the field. Report OD 150, HR Wallingford, Wallingford

Downward S, Taylor R (2007) An assessment of Spain's Programa AGUA and its implications for sustainable water management in the province of Almería, southeast Spain. J Environ Manage 82:277–289

European Commission (2002) Implementation of Council Directive 91/676/EEC concerning the protection of waters against pollution caused by nitrates from agricultural sources. Synthesis from year 2000 Member States reports, Report COM(2002) 407, Directorate-General for Environment, Office for Official Publications of the European Communities, Luxembourg

European Environment Agency (2003) Europe's water: An indicator-based assessment. Topic Report No 1, EEA, Copenhagen

European Environment Agency (2005a) European environmental outlook. EEA Report No. 4, EEA, Copenhagen

European Environment Agency (2005b) Sectoral use of water in regions of europe, EEA data Service, EEA, Copenhagen

Institut Français de L'Environnement (2005) Donn'ees essentielles de l'environnement. IFE, Orleans

Instituto Nacional de Estadística (INE) (2004) Estadísticas del Agua 2002. INE, Madrid

Instituto Nacional de Estadística (INE) (2005) Cuentas satélite del agua en España. INE, Madrid

Martínez Y, Albiac J (2004) Agricultural pollution control under Spanish and European environmental policies. Water Resour Res 40(10): doi:10.1029/2004WR003102

Martínez Y, Albiac J (2006) Nitrate pollution control under soil heterogeneity. Land Use Policy 23(4):521–532

Martínez L, Hernández N (2003) The role of groundwater in Spain's water policy. Water Int 28(3):313–320

Ministerio de Medio Ambiente (2000) Libro Blanco del Agua en España, Direcci'on General de Obras Hidráulicas y Calidad de las Aguas, Secretaría de Estado de Aguas y Costas. MIMAM, Madrid

Organization for Economic Co-operation and Development (OECD) (2007) OECD Environmental Data. Compendium 2004, OECD, Paris

Towards a Climate-Proof Netherlands

Michiel van Drunen, Aalt Leusink and Ralph Lasage

Introduction

There is no doubt about it: the climate is changing and the effects are now tangible and predictable. Scientific research has shown that even if we make significant reductions in greenhouse gas emissions (mitigation), climate change cannot be prevented. Which is why we have to adapt to make the effects of the changing climate acceptable: the Netherlands must be made climate-proof.

To stimulate climate-proofing, four ministries and the Climate *changes* Spatial Planning (CcSP), Living with Water (LmW) and Habiforum research programmes have established a National Programme on Adapting Spatial Planning to Climate Change (ARK, see Box 1). The core research questions examined by ARK are:

- What is the nature and scale of the observable and expected impacts of climate change for various themes and economic sectors?
- What spatial issues do they raise?
- How can we tackle these spatial issues?
- What dilemmas (technical, administrative, economic and social) will we face when trying to resolve these issues?

The Routeplanner is the scientific arm of ARK: the three research programmes, assisted by other research institutes, supply ARK with scientific information and insights on climate-proofing the spatial development of the Netherlands. This chapter summarises the outcome of phase 2 of Routeplanner, which took place in 2006. It consisted of a climate-proofing baseline assessment, a review (Quickscan) of knowledge gaps, formulation of adaptation strategies, a qualitative and quantitative assessment of adaptation options and an identification of case studies.

This chapter answers the following questions for the situation in the Netherlands, as far as current scientific understanding permits and reviewing the progress abroad (see Box 2):

A. Leusink (✉)
LOASYS, Management & Advice in Water and Environment, Groen van Prinstererlaan 34, 2271 En Voorburg, The Netherlands
e-mail: a.leusink@loasys.nl

- How will climate change affect the Netherlands?
- What are the consequences of climate change for the Netherlands?
- What must be done?
- When must we act?
- What examples are there of climate-proof strategies?
- What next?

Box 1 National Programme on Adapting Spatial Planning to Climate Change (ARK), mitigation and the role of developing countries

While recognising that climate change has a highly detrimental effect on developing countries, *ARK* focuses on making the Netherlands climate-proof. Some civil society organisations have problems with adaptation policies because they fear the government will then neglect mitigation. The corporate sector, on the other hand, wonders whether money spent on reducing emissions is money well spent. Adaptation and mitigation work on very different spatial and temporal scales. Adaptation delivers results within the short term and in specific places, even if other countries do little or nothing in response to climate change. Mitigation only has an effect in the long term and if other countries also cooperate. It aims at reducing the impacts of climate change across the whole world. If mitigation policy fails, no affordable adaptation options will be left open in the long run (after 2100). For that reason alone, a solid international mitigation policy is absolutely essential.

Box 2 What is happening in other countries?

In a 2005 report the *European Environment Agency (EEA)* (EEA, 2005) describes *Europe's* vulnerabilities to climate change and rising sea levels. Ecosystems, water resources, flood protection, forestry, agriculture and fisheries, public health, energy supply and tourism are all particularly vulnerable to climate change. The most vulnerable regions are the regions around the Mediterranean Sea and Central Europe (drought), the subarctic areas (ecosystems), the mountains (ecosystems, tourism) and the low-lying coastal areas (flood protection). Various countries have drawn up adaptation policies for specific sectors. *Norway* and *Finland* have policies for forestry, infrastructure and buildings; *Switzerland* and *Austria* for tourism and hydroelectric power; and the *Netherlands* and the *United Kingdom* for flood protection. Adaptation policies are being produced at a rapid rate.

In the *United Kingdom* climate change is considered to be an external risk on top of existing risks. An Adaptation Policy Framework for adaptation measures in response to climate change was drawn up in 2003. It contains directions for policymakers on the initial description of risks, assistance with defining the problem and specifying the objectives of the adaptation measures before a decision is made, and help with a review of the actions taken to monitor their effects (see also Box 3) (UK, 2006).

In 2004 *Denmark* concluded that it would experience limited direct impacts from climate change and is in a position to adapt to them. Little systematic attention has been given to the secondary effects of climate change, such as changes in recreational patterns, migration of refugees from threatened areas, and changes in the prices of agricultural goods. Most adaptation strategies are in the research stage and concrete policies have only been formulated for forestry (Danish Ministry of the Environment, 2005).

Finland published its national adaptation strategy in 2005. Its overall goal is the same as for the Netherlands: to make the country more resilient to the possible consequences of climate change. It identifies the following priority areas: incorporating the consequences of climate change and adaptation into mainstream sectoral policy, bringing climate change into decisions on long-term investments and coping with extreme weather events. Priority areas are agriculture and food production, forestry, fisheries, reindeer farming, water, biodiversity, industry, transport and communications, land use, communities, buildings and structures, public health, tourism and insurance (Ministry of Agriculture and Forestry, 2006).

The *United States* has not yet produced any policies specifically on adaptation, but the most vulnerable areas, sectors and issues have been identified (US Department of State, 2002). The *Australian* government has recently produced a decision framework for public authorities and companies to use for determining the need for policies on adapting to climate change. Their approach to climate adaptation is based on a risk analysis (Australian Greenhouse Office, 2006).

How Will Climate Change Affect the Netherlands?

In 2006 the Royal Netherlands Meteorological Institute (KNMI) published climate scenarios for the Netherlands for 2050 and 2100 (KNMI, 2006). For the first time in their analyses of the future climate they used a whole range of advanced global and regional climate models combined with information from time series of measured data, which allowed them to incorporate changes in air flow patterns in their models. Given the uncertainties about whether and how these flows are affected by the enhanced greenhouse effect, the KNMI decided to use two sets of climate scenarios: one set in which the flow patterns remain unchanged (current situation) and a second set in which the flow patterns do change. The latter are indicated by a '+' in Table 1.

The calculations for the climate scenarios with altered circulation patterns provide strong evidence for more frequent dry summers similar to those experienced in 1976 and 2003. Both sets consist of two scenarios. In the first scenario the average global temperature in 2050 is 1°C higher than in 1990, and in the second scenario it is 2°C

Table 1 Description of four KNMI climate scenarios

Code	Name	Description
G	Moderate	1°C temperature rise on earth in 2050 compared with 1990. No change in air flow patterns in Western Europe
G+	Moderate+	1°C temperature rise on earth in 2050 compared with 1990 + milder and wetter winters due to more westerly winds + warmer and drier summers due to more easterly winds
W	Warm	2°C temperature rise on earth in 2050 compared with 1990. No change in air flow patterns in Western Europe
W+	Warm+	2°C temperature rise on earth in 2050 compared with 1990 + milder and wetter winters due to more westerly winds + warmer and drier summers due to more easterly winds

Source: KNMI, 2006.

higher than in 1990. Extreme changes – such as those that would be caused by a reversal of ocean currents and which would cause widespread disruption to society – have not been included because the chances of such events occurring are low. According to current understanding, there is an 80% chance that the trends in the Dutch climate will be within the range covered by the four scenarios. This means that in 2050 there is an 80% chance that the average winter temperature will rise by between 0.9°C and 2.3°C and that the sea level will be 15–35 cm higher than in 1990 (see Table 2).

Table 2 Climate scenarios for 2050 for the Netherlands

2050 (compared to 1990)		G	G+	W	W+
Worldwide rise in temperature (°C)		+1	+1	+2	+2
Changes in air flow patterns in Western Europe		No	Yes	No	Yes
Winter	Average temperature (°C)	+0.9	+1.1	+1.8	+2.3
	Coldest winter day per year (°C)	+1.0	+1.5	+2.1	+2.9
	Average precipitation (%)	+4	+7	+7	+14
	10 day total rainfall exceeded once in 10 years (%)	+4	+6	+8	+12
	Highest average daily wind speed per year (%)	0	+2	−1	+4
Summer	Average temperature (°C)	+0.9	+1.4	+1.7	+2.8
	Warmest summer day per year (°C)	+1.0	+1.9	+2.1	+3.8
	Average precipitation (%)	+3	−10	+6	−19
	10 day total rainfall exceeded once in 10 years (%)	+13	+5	+27	+10
	Potential evaporation (%)	+3	+8	+7	+15
Sea level	Absolute rise (cm)	15–25	15–25	20–35	20–35

Source: KNMI, 2006.

Observations in recent years indicate that the average temperature in the Netherlands is rising faster than the global average and that extreme high temperatures are occurring more frequently. The warmer scenarios appear to describe the current situation best. The KNMI does not yet know whether these phenomena are a product of natural variability or a speeding up of climate change in the Netherlands. If the latter, the impacts will be more severe and the costs will be higher than those described below.

Table 2 lists the percentage changes from the climate in the baseline year 1990. To give an impression of what the change in climate could mean for the Netherlands, we compare the expected future Dutch climate with places in Europe which have a similar climate. The average changes in the seasons indicate that Dutch summers will in future be similar to those now found on the west coast of France around Bordeaux. Winter in Bordeaux is at present somewhat warmer than the expected Dutch winter climate, which will in future be more like that now found in the Po valley in northern Italy (Milan–Venice) (Kwadijk et al. 2006).

In the KNMI scenarios the trends in Table 2 continue beyond 2050. For 2100 they show an increase in average temperature of between 1.7°C and 5.6°C and a rise in sea level to between 35 cm and 85 cm above the 1990 level. Table 3 links the relevant climate factors to each theme and their main consequences, which are described in more detail in the next section. We use the following categories to describe the degree of uncertainty involved:

Very likely	More than 90% probability
Likely	66–90% probability
Medium likelihood	33–66% probability
Unlikely	10–33% probability
Very unlikely	Less than 10% probability

These categories are also used by the Intergovernmental Panel on Climate Change (IPCC) (IPCC, 2001). The percentages are based on expert judgements, taking account of the most recent results from research into changes in the climate and their effects. The target year is 2050.

What Are the Consequences of Climate Change for the Netherlands?

Climate-Proofing, Resistance and Resilience

The concept of climate-proofing is still in its infancy. From an engineering perspective, climate-proofing is the capacity of a system to continue to function well as the climate changes. It is a measure of the range within which the system, such as an ecosystem, a socio-economic system or a technological system, continues to

Table 3 Effects of climate change on different systems

	Increase in average temperature (vl)	Increase in number of heat waves, summer droughts (vl)	Increase in intensity of summer precipitation (l)	Increase in winter precipitation (l)	Increase in wind velocity during storm events (md)	Sea level rise (vl)
Water			More frequent flooding of urban areas (l)	Increase in peak discharges from rivers thus raising chance of flooding (vl) Increase in flooding of rural areas (vl)	Worsening of coastal erosion and flooding by the sea (vl)	Intensification of coastal erosion and flooding by the sea (vl) Rise in water levels in low-lying Netherlands and an increase in chance of flooding (vl)
Nature, agriculture	Extension of the growing season (vl)	Decrease in surface and groundwater levels, salinity (vl)		Increase in surface water levels (vl)		
Energy	Decline in natural gas consumption (vl) Increase in electricity consumption (vl)	Increase in frequency of cooling water constraints (vl)			Increase in damages to high-voltage lines due to extreme storms (md)	
Transport	Fewer problems due to extreme winter conditions (vl) Lower chance of constraints on water transport capacity due to ice cover (l)	Reduced transport capacity of river vessels due to low water levels (vl)	Increased hindrance of heavy rainfall (vl)	Reduced transport capacity of river vessels due to high water levels (vl)	Increase in damages to vehicles due to extreme storms (md)	

Table 3 (continued)

	Increase in average temperature (vl)	Increase in number of heat waves, summer droughts (vl)	Increase in intensity of summer precipitation (l)	Increase in winter precipitation (l)	Increase in wind velocity during storm events (md)	Sea level rise (vl)
Housing and infrastructure	Fewer occasions when roads need to be gritted or salted (vl)	Increased occurrence of melting road surfaces due to heat (vl)	Increase in damages to infrastructure and buildings due to flooding (vl)	Increase in flooding (vl) Increase in river discharge (vl) Increase in damages to buildings due to flooding (vl)	Increase in damage to oil rigs, roads, bridges, buildings and vehicles due to extreme storms (vl)	
Public health	Increase in Lyme disease, allergies (md)	Drop in air quality (vl)				
Recreation	Rise in recreation (l) Limitations in water recreation (l)	Decline in bathing water quality (vl)			Worsening of dune and beach erosion (vl)	Worsening of dune and beach erosion (vl)

Source: Kwadijk et al. (2006). vl = very likely, l = likely, md = medium likelihood

function 'normally'. Climate-proofing can be considered to be the result of two other, more specific system characteristics: resistance and resilience. Here, *resistance* refers to the capacity to withstand an external pressure without reacting, and *resilience* is the capacity to absorb the pressure and recover quickly once it has gone (Gouldby et al. 2005; De Bruijn, 2005).

The Dutch coastal defences, for example, have a high resistance and a low resilience to climate change: they can hold back a limited rise in sea level and more powerful storms, but once breached they would take a long time to recover to their previous state. The cooling systems in power stations, on the other hand, have a low resistance and high resilience: they will consistently fail when the temperature of the river water rises, but as soon as the temperature falls they function normally again.

Risks and Opportunities

The above examples already show that it will not be easy to determine whether the Netherlands as a whole is climate-proof: it depends on which 'system' you look at and how you define resistance and resilience in these systems. For example, the resistance of the coastal defences can be expressed as the probability of flooding. This is the traditional approach, which underlies the Delta plan. But to compare how climate-proof the coastal defences are against other systems we need to know the risk of flooding. A commonly used method to calculate this is the probability of failure times the severity of damage, but this rational approach is often not very satisfactory because the public and experts can hold different opinions about the seriousness and scale of risks (and experts can also disagree among themselves). Participatory methods can be used to bring social values, norms and convictions into the process of establishing the risks (Osinga en Kabat, 2006; Veraart et al. 2006).

Besides the engineering approach to climate-proofing, the Routeplanner introduces the concept of 'climate-proof land use'. This is a state in which climate-related risks are reduced to a socially and economically acceptable level by a series of combined technological and social adaptation measures and low-emission land use. Climate change and the knowledge gained in the process can then generate socio-economic opportunities in the form of advanced social, technological and institutional innovations (Veraart et al. 2006).

Socio-economic Scenarios

Determining what is climate-proof is also complicated by the fact that the effects of climate change will (mainly) be felt in the future. As Dutch society is continuously changing there is little sense in projecting future climate effects onto the current socio-economic situation. The Routeplanner gets around this problem by using socio-economic scenarios in the analysis. These scenarios were developed in The Future of the Dutch Natural and Built Environment (Welvaart en Leefomgeving, WLO) by the Netherlands Bureau for Economic Policy Analysis (CPB), the

Netherlands Environmental Assessment Agency (MNP) and the Netherlands Institute for Spatial Research (RPB) (WLO, 2006). They envisage that, while climate change will affect Dutch agriculture, its impacts will be marginal compared with the consequences of future European agricultural policy. These WLO scenarios also respond differently to the opportunities and threats of climate change, for example, in relation to flood risks. Both the KNMI and WLO scenarios assume stable policy trends in which there is no place for significant emission reductions and new adaptation policy.

Water

The main effects of climate change on the water system are (no measurement):

- A rise in sea level, thus raising the likelihood of coastal erosion and flooding (very likely)
- An increase in peak discharges from the rivers in the winter, thus raising the likelihood of flooding (very likely)
- An increase in water levels in the IJsselmeer area and the main inland waterways of Zuid-Holland and Zeeland, thus raising the likelihood of flooding (very likely)
- An increase in flooding during winter in rural areas (very likely)
- More frequent flooding in urban areas (likely)
- An increase in household consumption of drinking water in summer (very likely)
- Greater penetration of saline water into surface water bodies (very likely)
- Salination of groundwater resources (unknown)
- Lowered groundwater tables (unknown)

Over the past 100 years the sea level has risen by about 20 cm. The rising sea level leads to erosion of sand from the coast and reduces the safety levels of sites directly on the coast. The climate scenarios also predict higher wind speeds, although the increase is small and lies within the current variability in wind speed from year to year.

The expected higher precipitation in winter will feed into higher discharges through the flood basin of the Rhine and Meuse. In the Rhine this effect will be reinforced by the rise in temperature, which will cause more rain and less snow to fall in the Alps during the winter. The combined effect of these two trends almost certainly means higher winter discharges in the Rhine. The likelihood of flooding from the river will rise accordingly, with the greatest likelihood in the lower reaches of the rivers in the province of Zuid-Holland, downstream from around Zaltbommel or Vianen. Here the river levels are a product of the discharge through the river and the – higher – water levels at sea.

The higher winter discharges of the Rhine will be accompanied by greater volumes of water flowing through the river IJssel to the IJsselmeer, the lake created by the enclosure of the former Zuider Zee. Drainage from the IJsselmeer will become more difficult because water levels in the Wadden Sea will also rise. A rise in winter precipitation in the polders will lead to more frequent flooding in winter; land will be under water more frequently or for longer periods. As the sewage systems were designed to cope with less violent downpours, heavier summer storms will mean more flooding in urban areas.

It is very likely that average demand for drinking water will rise by a few per cent as the temperature rises. More frequent heat waves will cause more frequent peaks in the demand for water. Some surface water abstractions (for example at Ridderkerk) will experience more frequent periods when the water is too saline for treatment and the intake has to be temporarily shut down. This usually happens in summer when the demand for drinking water is highest. Preparing drinking water from poorer quality or even brackish water is technically possible, but very expensive.

A higher risk of flooding is not just a reflection of the likelihood of it happening; as a first approximation at least, risk is the probability multiplied by the damage. The WLO scenarios assume the potential for damage is growing faster than the threat from climate change because the capital investments in areas susceptible to flooding, in relative terms, are rising faster.

Nature

The main effects of climate change on the natural system are (no measurement):

- An increase in surface water levels in winter (very likely)
- A decrease in surface water and groundwater levels in summer (very likely)
- Wetter winters (very likely)
- Drier summers (very likely)

Climate change will allow some plant and animal species from warmer, more southerly regions to become established in the Netherlands. Examples include the oak processionary caterpillar and several lichens. Species with limited means of dispersal will be restricted in their range by the fragmentation of ecosystems and habitats.

Plant growing and flowering periods and bird breeding times have shifted in response to climate change and food chains are being disrupted in the process. Hydrological changes in groundwater and surface water and temperature changes are putting ecosystems such as forests, grasslands, coasts and fenland under increasing pressure. Our aquatic and wet terrestrial ecosystems, such as stream and river systems, wetlands, wet heath and raised bog, are particularly sensitive to extremes in the weather. Rising sea levels affect the coastal ecosystems by eroding the coastal dunes and flooding salt meadows.

Climate is the overriding parameter for natural systems, and climate change aggravates the effects of habitat fragmentation and water table drawdown. Although the WLO scenarios presume an improvement in water quality and a more robust National Ecological Network, these cannot compensate for the negative consequences of climate change.

Agriculture

The main effects of climate change on agriculture are (Kwadijk et al. 2006):

- Greater disruption in winter from high water levels and flooding in the rural areas of the lower-lying north and west of the Netherlands (very likely)
- Extension of the growing season (very likely)
- More frequent and longer lasting soil water deficits during the summer (very likely)
- An increase in the likelihood of brackish groundwater seepage (very likely)

Changes in the climate will generally improve the average climatic conditions for farming in the Netherlands. Higher temperatures mean longer growing seasons and thus higher potential crop yields. Energy costs for heating greenhouses will fall because of higher winter temperatures. The four WLO scenarios paint a set of very different prospects for agriculture: the sector is sensitive to the main variables in the scenarios – international cooperation and market liberalisation – and possibilities for expansion in the agricultural sector are limited by environmental policy and animal welfare policies. Markets are also becoming saturated and the cultivated area in the Netherlands is shrinking. Market liberalisation (in two of the four scenarios) accounts for an expansion in dairy farming, but a contraction of arable farming and intensive livestock farming. Dutch agriculture is largely independent of the climate and can react flexibly to changing climatic conditions; smaller yields in dry years will often be compensated by higher prices. Extreme weather will have only a limited effect on the economic success of the sector.

Energy

The consequences of climate change for the energy sector are (Kwadijk et al. 2006):

- A decline in natural gas consumption in winter (very likely)
- An increase in electricity consumption in summer (very likely)
- An increase in the frequency of cooling water constraints (very likely)

It is very likely that natural gas consumption in winter will decline because less will be needed for space heating in homes, offices and greenhouses. However, the reduction in CO_2 emissions will be largely cancelled out by an increase in electricity consumption for air conditioning in the summer.

Power stations extract surface water for cooling and so higher surface water temperatures are a major problem for them. If the intake water is warmer than 23°C the discharge water is so hot that it causes oxygen deficiencies downstream, which is disastrous for river ecosystems. Cooling water is already in short supply during dry years, sometimes preventing the power stations at Bergum, Diemen, along the Amsterdam-Rhine Canal and along the North Sea Canal from operating at full capacity. Further warming of the water will make cooling water constraints more frequent and electricity production will fall.

The degree to which the electricity supply system can continue to operate under extreme conditions is determined partly by existing overcapacity. In recent years overcapacity in the privatised energy sector has declined as efficiency has gained in importance. Some overcapacity guarantees greater security of supply but is expensive to maintain, which is reason enough to abandon it on commercial grounds. Energy demand varies considerably in the four scenarios, but fossil fuels remain important in all of them. All the scenarios assume an increase in electricity demand, but here too the differences are wide: the lowest is +16% and the highest is +123% compared with 2002.

Transport

The consequences of climate change for the transport sector are (Kwadijk et al. 2006):

- Increasing problems of heavy rainfall (very likely)
- More corrosion of vehicles (likely)
- More damage to vehicles from extreme storms (medium likelihood)
- Increased constraints on transport from air pollution (unknown)
- Fewer problems due to extreme winter conditions (very likely)
- Fewer accidents due to fewer days with frost (unknown)
- Fewer delays to journeys because of snow (very likely)
- Reduced transport capacity and loading capacity of river vessels due to periods when water levels are too high or too low (very likely)
- Lower chance of constraints on water transport capacity from ice cover (likely)

Driving conditions will be affected by extreme downpours in the summer and more rain in the winter. On the other hand, there will be fewer disruptions because of extreme winter weather (snow and icy conditions). An increase in the number of extreme rainstorms may cause temporary hold-ups to air traffic. Metal corrosion will become more of a problem owing to the higher temperatures and will probably force a shorter write-off period for vehicles.

By attracting many people to the coast, heat waves will be accompanied by traffic congestion. As people suffering from heat stress are less alert, the likelihood of accidents will increase. Heat waves are also often accompanied by a sharp rise in air pollution, but it is still unclear how far traffic will be affected by measures taken to curb air pollution.

It is very likely that periods with low river levels will become more common in summer and periods of high river levels will be more frequent in winter. Both will limit the number of vessels and/or prevent loading capacity being used to the maximum. Ice will probably be less of a problem. Extreme weather conditions will have incidental impacts on the transport sector. Assuming that weather conditions in the Netherlands will become like those in Southwest France, no major impacts

Towards a Climate-Proof Netherlands

on road or rail traffic are to be expected. The relatively short depreciation periods for investments in the road haulage sector allow it to react flexibly to climate change. Investments in transport by rail and waterway, however, require more time and the replacement periods of materials are much longer, making them more vulnerable. The effects of low river discharges could become an important factor in water transport.

Other developments are much more important for the performance of the transport sector, such as economic growth, energy prices and the available infrastructure. The WLO scenarios show increasing congestion on the Dutch roads by 2020; in three of the scenarios congestion eases thereafter and in the other one it doubles by 2040.

Housing and Infrastructure

The consequences of climate change for housing and flood protection infrastructure are:

- Increase in corrosion due to higher precipitation and higher temperatures (likely)
- Increase in damage to oil rigs, high voltage transmission lines, roads, bridges and buildings from extreme storms (medium likelihood)
- Greater occurrence of melting road surfaces in hot temperatures (very likely)
- Fewer occasions when roads need to be gritted or salted (very likely)
- Reduction in damage to rail tracks and roads by frost and salt/grit, and fewer inspections required (very likely)
- Reduction in ice accretion on wind turbines (very likely)
- Increased damage to buildings from flooding (very likely)
- Reduction in navigability of river arms during the summer (very likely)
- More frequent obstructions to shipping during high water in winter (very likely)

Extreme storms can damage buildings and infrastructure directly or cause damage indirectly through falling trees and the like. High-voltage transmission lines and other overhead lines are particularly vulnerable, and some bridges and dikes will have to be closed during powerful storms. The Stern Review Report (Box 3) states that the economic costs of storms and floods could be very high.

Box 3 Main conclusions of the Stern Report

The Stern Review Report to the British Government reviews the impacts of climate change, adaptation options and the economic costs of transforming society into one that is much less dependent on fossil fuels. The report was published just after the Routeplanner studies had been completed, but contains some identical conclusions:

- There is still time to prevent the worst effects of climate change, if we act now.
- Climate change can have serious consequences for economic growth and development; for example, without adaptation measures the costs of flooding in the UK will rise from 0.1% to 0.2–0.4% of GDP if it becomes 3–4°C warmer in summer. The additional costs of climate-proof infrastructure and buildings in OECD countries are € 11–110 billion per year (0.05–0.5% of GDP).
- The costs of stabilising the climate are considerable, but not insurmountable: the annual cost of stabilising greenhouse gas concentrations at 500–550 ppm CO_2 equivalents is about 1% of GDP in 2050. Further delay in taking action is risky and will entail higher costs in the end.
- All countries must take climate measures, but these should not frustrate the growth aspirations of both poor and rich nations.
- Many measures can be taken to reduce emissions, but targeted policies are needed to get them off the ground.
- The response to climate changes must be international and based on agreement about the long-term objectives and the principles underlying the measures to be taken.

Source: Stern Review Report (Stern, 2006).

More precipitation combined with higher temperatures will accelerate corrosion of viaducts, bridges and other infrastructure, and inspections and maintenance work will be needed more often. More frequent heat waves will damage roads because of melting asphalt. Salt and grit spreading in winter will be needed on fewer occasions, resulting in less damage from frost and salt. Ice accretion on wind turbines will occur less frequently.

A decline in the discharges of the main rivers during the summer will lead to lower water levels, reducing navigability of the river channels and limiting load draughts during dry periods. An increase in the winter discharges will cause frequent obstruction to shipping because of restricted headways under bridges.

Damage to buildings may increase as a result of extreme storms and flooding and urban areas could become 'heat islands' in summer. Heat stress will feed demand for 'intelligent' buildings built to stay cooler in summer. Existing buildings have a depreciation period of 40–50 years, and retrofitting these buildings is an expensive option. But given the rate of climate change, the changing functional demands being made on buildings (such as more comfort) are of greater influence than the demands made on the building by the changing climate.

Investments in infrastructure are large and can have long depreciation periods. Once it is built, infrastructure is highly persistent. Weather conditions are an important factor in the usability of road and rail and the regularity of maintenance work. Compared with the change in the levels of use (increased traffic, heavier vehicles), climate change makes little contribution to wear and tear, proceeding as it does at a slow pace compared with the frequency of regular maintenance work of dry infrastructure.

Public Health

The direct consequences of climate change for public health are (no measurement):

- An increase in mortality during summer (very likely)
- A reduction in mortality during winter (medium likelihood)
- An increase in mortality from flooding and high water levels (unlikely)
- An increase in stress caused by more frequent flooding and high water levels (very likely)
- Higher mortality from storms (medium likelihood)

Indirect consequences are

- Vector-transmitted diseases
  - Increase in malaria (unlikely)
  - Increase in Lyme disease (medium likelihood)
- Diseases linked to air quality
  - Increase in summer smog (ozone and particulates) (likely)
  - Reduction in winter smog (particulates) (medium likelihood)
- Allergies
  - Increase in hay fever (likely)
  - Increase in house mite allergy (unknown)
- Increase in water-related diseases (medium likelihood)
- Increase in food-related diseases (unlikely)
- Increased exposure to UV-related disorders (medium likelihood)

The direct consequences of climate change are an increase in the chances of mortality from heat stress and a decrease in the chances of mortality from extreme cold. Flooding and high water also cause stress. During heat waves the incidence of heat stress temporarily raises the mortality rate about 15% above the rate during a comparable period without a heat wave. These extra fatalities include older people who die a few weeks sooner because of the heat and some who probably die as a result of the poorer air quality that usually accompanies a heat wave. Mortality rates also rise during periods of extreme cold, but the number of cold periods is expected to decrease (medium likelihood). The air pollution (particulates) frequently found during extremely cold periods will also be reduced.

Many of the indirect effects are related to changes in behaviour: people are expected to go outside more often and for longer because it will become warmer on average and they will spend more time on outdoor leisure and recreation activities. Exposure to UV radiation, air pollution and pollen, waterborne diseases (cyanobacteria, amoebae) and Lyme disease will increase as a result. Climate change in the Netherlands will increase the chances of contracting Lyme disease (medium likelihood) and lead to more cyanobacteria in surface waters (medium likelihood). House mite allergy may become more common because winters will probably be wetter and so the indoor climate will be moister as well. The ozone layer above the Netherlands will probably recover more quickly because of climate change, reducing exposure to UV radiation. Although food goes off quicker at higher temperatures, this is not considered to be a risk because of the high food safety standards in the Netherlands.

In all four WLO scenarios, the health-care sector makes up a much greater proportion of the economy than at present. The reasons for this include the ageing population, advances in medical technology and social and cultural factors such as the expectations of – relatively rich – patients. Although climate change has an influence on diseases, and thus on human health, other factors have a much greater impact. Examples include frequent travel, which makes it much easier for diseases to spread, as well as infectious diseases, quality of the indoor environment, lifestyle and eating patterns (obesity and cardiovascular disease).

Recreation

The consequences of climate change for the recreation sector are (no measurement):

- Worsening of dune and beach erosion (very likely)
- Restrictions on water-based recreation, such as reduced navigability and more delays at bridges and locks (likely)
- Decline in bathing water quality (medium likelihood)
- Increase in numbers of day trips (unknown)
- Rise in number of foreign tourists (unknown)

Rising sea levels and more severe storms will increase erosion of beaches and dunes. Beach nourishment or recharging with imported sand will be needed on a large scale to maintain beach width, which will make coastal tourism dependent on successful beach recharging schemes.

The water sports sector will probably face a reduction in water quality, poorer fish stocks and more delays at bridges and locks because of lower river discharges in summer. Bathing water quality has the greatest impact on recreation, but the positive effect of good weather will be much bigger than the negative effects of lower bathing water quality. Depending on the climate scenario, net spending in the recreation sector will rise by between 1% and 6%. However, no account has been taken of any changes in leisure and recreation behaviour.

European studies show that in the months of June, July and August the temperature in the traditional holiday regions in the Mediterranean will be too high for many tourists. In the more temperate climates, on the other hand, conditions will become more favourable. The numbers of foreign tourists coming to the Netherlands may rise, and more people may stay in the Netherlands for their holiday.

What Must Be Done?

The previous chapter has shown that a range of systems in the Netherlands are sensitive to climate change. The Routeplanner describes no less than 96 adaptation options to reduce these vulnerabilities (van Ierland et al. 2006). The options were evaluated against five criteria: importance, urgency, no regret (also favourable without climate change), side effects and mitigation effects (reduced amounts of greenhouse gases). The criteria were scored on a scale from 1 to 5, 5 being the highest score. Table 4 shows the 44 options with the highest weighted total, plus two options which received a 5 for urgency or importance. The weight for each criterion is given between brackets. Radical options, such as building artificial reefs in the sea and abandoning the lower-lying areas of the Netherlands, are not included in the table because they received low marks for all the criteria, at 85 and 95 on the list.

The table also shows the complexity of the issues measured as the weighted sum of the technical, social and institutional complexity: the higher the score, the lower the feasibility of the option. Social complexity depends on the number of stakeholders, the diversity of the values they hold and the level of resistance to the proposed measure. Institutional complexity consists of the conflicts between administrative regulations, the consequences for institutions, the degree of cooperation required and the changes to current administrative arrangements. The different formatting in numbers indicates the themes: bold for water, italic for nature, bold-italic for housing and infrastructure, underline for energy, underline-italic for agriculture and underline-bold for public health.

A conspicuous feature of the table is the large number of options for the water theme (37%) and the low number for public health (2%). The other themes each make up about 15% of the total. This is more or less in line with the analysis of climate resilience per theme in the previous chapter.

The high scores for complexity obtained by many of the options in Table 4 show that the barriers to implementation are high. Many of these barriers are institutional in nature and the Routeplanner recommends finding new, temporary and flexible institutional arrangements that permit an integrated and coordinated approach. In many cases, the social complexity is high because the options involve a high number of stakeholders. Realising such options will require a higher sense of urgency among many of these stakeholders. Significant input from stakeholders (farmers, fishers, residents) in the decision-making process creates public support and makes it easier to implement the measures. In technical terms, the measures are generally not very complex.

Table 4 Ranking of adaptation options in the Routeplanner scored for the five criteria. The complexity of their introduction, which was not included in weighting, is also included

Adaptation options	Sector	Weighted sum	Importance (40%)	Urgency (20%)	No regret (15%)	Ancillary benefits (15%)	Mitigation effect (10%)	Complexity
More space for water	w	**4.9**	**5**	**5**	**5**	**5**	4	**4.4**
Risk-based allocation policy	w	**4.9**	**5**	**5**	**5**	**5**	4	**4.4**
Risk management as basic strategy	w	**4.9**	**5**	**5**	**5**	**5**	4	3.2
New institutional alliances	w	**4.9**	**5**	**5**	**5**	4	**5**	**4.0**
Integrated nature and water management	n	4.9	5	5	5	5	4	4.2
Integrated coastal zone management	n	4.9	5	5	5	5	4	4.2
Make existing and new cities robust – avoid 'heat islands', provide for sufficient cooling capacity	u	**4.8**	**5**	**5**	4	**5**	4	3.0
Construct buildings differently in such a way that there is less need for air conditioning/heating	i	4.7	5	4	5	4	5	2.6
Change modes of transport and develop more intelligent infrastructure	i	4.7	5	5	4	4	5	4.0
Evacuation plans	w	**4.5**	**5**	**5**	**5**	3	3	**4.0**
Design and implementation of ecological networks (The National Ecological Network – NEN)	n	4.5	4	5	5	5	4	3.6
Design spatial planning – construct new housing and infrastructure	u	4.5	5	5	4	3	4	4.0

Table 4 (continued)

Adaptation options	Sector	Weighted sum	Importance (40%)	Urgency (20%)	No regret (15%)	Ancillary benefits (15%)	Mitigation effect (10%)	Complexity
Design houses with good climate conditions (control) – 'low energy'	u	**4.5**	**5**	**4**	**5**	**3**	**5**	**2.4**
Development of more 'intelligent' infrastructure that can serve as early warning indicator	i	<u>4.5</u>	<u>5</u>	<u>4</u>	<u>4</u>	<u>4</u>	<u>5</u>	<u>2.6</u>
Increasing genetic and species diversity in forests	n	*4.4*	*4*	*5*	*5*	*4*	*4*	*2.8*
Freshwater storage to flush brackish water out during dry periods	w	**4.3**	**5**	**3**	**5**	**4**	**3**	**4.0**
Maintain higher water table to prevent salt water intrusion	w	**4.3**	**5**	**3**	**5**	**4**	**3**	**3.8**
Afforestation and mix of tree species	n	*4.3*	*4*	*4*	*5*	*4*	*5*	*2.3*
Educational programmes	n	*4.3*	*4*	*4*	*4*	*5*	*5*	*3.0*
Widening the coastal defence area (in combination with urbanisation and nature)	w	**4.2**	**5**	**4**	**3**	**4**	**3**	**2.3**
Re-enforcement of dikes and dams, including 'weak spots'	w	**4.2**	**5**	**5**	**3**	**3**	**3**	**2.3**
Creating public awareness	w	**4.2**	**4**	**5**	**4**	**3**	**5**	**2.7**
Revision of sewer system	u	**4.2**	**5**	**4**	**4**	**3**	**3**	**2.7**
New design of large infrastructure	u	*4.2*	*5*	*4*	*4*	*3*	*3*	*3.7*

Table 4 (continued)

Adaptation options	Sector	Weighted sum	Importance (40%)	Urgency (20%)	No regret (15%)	Ancillary benefits (15%)	Mitigation effect (10%)	Complexity
Introduction of ecosystem management in fisheries	a	4.2	4	4	5	5	3	2.7
Monitoring nature, interpreting changes and informing	n	4.1	4	5	5	3	3	3.0
Spatial planning of locations for powerplants (nuclear in particular)	w	4.0	5	4	3	3	3	3.3
Relocation of freshwater intake points	w	4.0	5	4	3	3	3	3.3
Adapted forms of building and construction	w	4.0	4	4	3	5	4	3.3
Adaptation of highways, secondary dikes to create compartments	w	4.0	5	4	5	1	3	3.7
Stimulate economic activity in other parts (eastern and northern) of the Netherlands	w	4.0	5	5	3	3	1	2.7
Options for water storage and retention in or near city areas	u	4.0	4	4	4	4	4	2.7
Lowering the discount factor for project appraisal	i	4.0	4	5	3	3	5	4.0
Development of cooling towers	i	4.0	5	3	3	4	3	4.0
Increase standards for buildings so as to make them more robust to increased wind speeds	i	3.9	5	4	4	1	3	4.0

Table 4 (continued)

Adaptation options	Sector	Weighted sum	Importance (40%)	Urgency (20%)	No regret (15%)	Ancillary benefits (15%)	Mitigation effect (10%)	Complexity
Introduction of southern provenances of tree species and drought-resistant species	a	3.9	4	3	5	4	3	3.7
Acceptation of changes in species composition in forests	a	3.9	4	3	5	4	3	2.7
Changes in farming systems	a	3.8	4	3	4	4	4	2.0
Adjustment of forest management	n	3.7	4	3	5	3	3	4.0
Emergency systems revision for tunnels and subways	u	3.7	5	3	4	1	3	4.3
Constructing more stable overhead electricity transmission poles	i	3.7	5	3	4	1	3	4.3
Improvement of vessels	i	3.7	4	4	3	4	2	4.7
Water storage on farmland	a	3.7	3	4	5	4	3	4.3
Higher water level IJsselmeer	w	3.6	5	2	3	3	3	4.3
Increase sand suppletions along coast	w	3.5	5	3	3	1	3	3.7
Improvement of health care for climate-related diseases	h	3.4	4	5	3	1	2	2.0

Source: van Ierland et al. (2007). Explanation of sector: w = water, n = nature, u = urban, i = infrastructure, a = agriculture/fishery, h = health care

A correlation can be drawn between ranking and complexity: many of the important and urgent options are hard to implement. Examples include more room for water, spatial planning informed by risks and new institutional alliances for water management. Nevertheless, some important and urgent options are relatively easy to implement, such as nature education and many of the more technical measures.

When calculating the benefits of the adaptation options it is important to choose a good reference point. Table 5 sets out the possibilities, from which we can determine the costs and benefits as follows (van Ierland et al. 2007):

- The cost of climate change is the difference between the wealth of society at present and the unadapted society: $W(A0, T1) - W(A0, T0)$
- The net benefit of adaptation is the difference between the wealth of the adapted society and the unadapted society: $W(A1, T1) - W(A0, T1)$
- The cost of climate change after adaptation is the difference between the wealth of the adapted society and present-day society: $W(A1, T1) - W(A0, T0)$

Relatively little is known about the costs and especially the benefits of the adaptation options investigated. As can be seen from Table 6, data are available for only 17 of the 46 options and the benefits are monetarised for just 7 options. Uncertainty levels are very high where the costs and benefits have been monetarised. The costs cannot be added up because a choice for option X could mean that option Y would become unnecessary, or partly redundant. Although the Routeplanner provides an overview of costs and benefits for the first time, much more research is still needed. The exact costs and benefits of many of the adaptation options can only be determined once the locations of the measures to be taken are known.

The costs of the options differ widely: the most expensive options are for water retention measures to 2050, costing more than € 19 billion. For some options, for example, forestry and preventing heat islands, the benefits are higher than the costs, which makes them attractive from an economic point of view. Most of the options with the highest costs are connected with flood protection, but here it is not very clear what proportion of the costs are due to climate change and how much is due,

Table 5 Costs and benefits of adaptation

Adaptation type	Climate does not change (T0)	Climate changes (T1)
Adaptation to current climate (A0)	Current society (A0, T0)	Society did not adapt to the changed climate (A0, T1)
Adaptation to changed climate (A1)	Society is 'adapted' but climate did not change (A1, T0)	Society is adapted to changed climate (A1, T1)

Source: adapted from the Stern Review Report (Stern, 2006).

Table 6 Indicative costs and benefits of adaptation options (in million euros) at 2006

Adaptation options	Net present value cost (million €)	Net present value benefits (million €)
More space for water		
– Regional water system	19,000	Unknown
– Improving river capacity	>7,000	Unknown
Risk-based allocation policy	0–10	Unknown
Make existing and new cities robust – avoid 'heat islands', provide for sufficient cooling capacity	65 €/m^2	>2200 €/m^2
Construct buildings differently in such a way that there is less need for air conditioning/heating	23,000	Unknown
Design and implementation of ecological networks (The National Ecological Network – NEN)	7,000	>7000
Increasing genetic and species diversity in forests	0.43/ha	>0.43/ha
Widening the coastal defence area (in combination with urbanisation and nature)	1,000	Unknown
Re-enforcement of dikes and dams, including 'weak spots'	>5,000	Unknown
Revision of sewer system	3,000–5,000	Unknown
Monitoring nature, interpreting changes and informing	340	>340
Relocation of freshwater intake points	50–100	Unknown
Options for water storage and retention in or near city areas	3,300	Unknown
Lowering the discount factor for project appraisal	0	Unknown
Development of cooling towers	275–500	6.6–11
Water storage on farmland	15–50	Unknown
Higher water level IJsselmeer	>500	Unknown
Increase sand suppletions along coast ≥ minimal	750–1,500	Unknown

Source: van Ierland et al. (2007)

for example, to changing safety standards. Other expensive adaptation options are construction of climate-proof buildings and upgrading sewer systems.

The costs and benefits are expressed as net cash value at 2006. A discount rate of 4% has been applied to reflect the fact that costs are more attractive the further in the future they will actually be incurred. Benefits, on the other hand, are preferred immediately. Higher net values are assigned to determined benefits realised in the future when the discount rate is lower. Lowering the discount rate, therefore, stimulates investment in benefits to be obtained in the (distant) future. This is often the case in climate adaptation options, which is why lowering the discount rate is also included as an adaptation option.

When Must We Act?

Climate change could cause considerable damage in the Netherlands. The Stern Review Report (Box 3) confirms that the effects of climate change could cause widespread disruption to society and that action must be taken soon. Stern recommends planning adaptation options for areas where high risks are expected and where costs are low. Adaptation will be an autonomous process, but policies will have to be put in place for many options, as indicated in Table 7. Good adaptation policy requires planning in advance and a choice of robust measures that are practicable in various climate scenarios and which could be adjusted at a later stage if required. Because many of the options are expensive, financing will require very careful planning. Table 4 shows that the water management options in particular are urgent and important, as well as the construction of climate-proof buildings and preventing heat islands in urban areas.

Deciding on adaptation measures with a spatial dimension is a complex process. Adaptation to climate change throws up complex challenges on four fronts (de Pater and van Drunen, 2006):

- Various sectors will have to work on efficient and effective solutions, with greater emphasis on development and change than on maintaining existing land uses
- The solutions, in different places and at various scales, must provide an effective answer on the scale at which climate change is operating
- Costs, benefits and risks must be fairly distributed between present and future generations and between countries
- Costs, benefits and risks must be fairly distributed between government, public and private organisations and citizens

In a time of administrative decentralisation and internationalisation, new strategies will have to be developed that can catalyse more decisive management. Integrated area development is one of them. Central government considers it one of its tasks to promote the uptake of adaptation in all area development projects and the development of appealing, wide-ranging development visions and policy tools. Improving the design and problem-solving capacities of the actors is also important.

Table 7 Examples of adaptation options

	Autonomous	Policy
Short term	Short-term adjustments, like planting date crops; sharing of risks, through for instance insurances	Development of further knowledge on climate risks, improve emergency response
Long term	Invest in climate resilience as future effects are clear and benefits large, for instance local irrigation	Invest in infrastructure; avoid the effects, for instance through spatial planning policy

Source: Stern Review Report (Stern, 2006)

Obstructive regulations should be dispensed with and replaced by regulations that encourage collaboration between different sectors and areas (hotspots).

Because institutional and social complexity is high, implementation of many adaptation options should involve participation by stakeholders, such as farmers and other citizens. The sense of urgency among many stakeholders will have to improve, though, if the participatory process is going to be successful (van Ierland, 2006). This underlines the importance of educational and awareness-raising programmes, which can also be considered to adaptation options.

Case studies that take an integrated view of an area can give us a better feeling for the interaction between the measures taken in that area and the administrative processes required to manage the often radical interventions. We need to make a quick start. Not only is the climate already changing, but we still have little experience with such integrated approaches. The lessons from the case studies will form a necessary step towards the successful implementation of administratively more complex adaptation options. Some examples of the case studies are described in the next section.

What Examples Are There of Climate-Proof Strategies?

After holding several workshops and interviews, the research programmes Climate *changes* Spatial Planning (CcSP), Living with Water (LmW) and Habiforum drew up a list of 39 potential demonstration projects for the Routeplanner (hotspots). The demonstration projects were then ranked according to criteria for the following facets:

- Policy: opportunities and constraints in the field of climate change and various themes
- Support: supported by several administrative tiers
- Communication: appealing to a wide public; effects of climate change are clarified
- Action perspective: without climate change the project would have been interpreted differently

In addition to the criteria mentioned above, the final list of prioritised demonstration projects has to reflect a balanced geographical distribution and thematic coverage. Three of the top five projects are described below. They are not intended to be representative, but provide a good indication of the opportunities and threats resulting from climate change in three very different areas.

Biesbosch

The Biesbosch is located between the southern edge of the Randstad and the Brabant city ring. This freshwater tidal zone performs important functions for a broad hinterland: power generation, recreation, shipping, water retention and drinking water supply. It has proved very difficult to achieve a sustainable water management regime because the region is one of the main nodes in the national water management system.

In the short run, the Rhine and Meuse, which run through the area, will have to be given room to expand. The need for this was amply demonstrated during the high water levels of 1995. The dikes along two river channels in the lower reaches of the Rhine and Meuse, the Beneden-Merwede and Beneden-Waal, just held because the sea was quiet at the time and water could be discharged continuously. The polders around the Haringvliet inlet were also badly hit during the extreme rainfalls in 1998, when they proved to have far too little temporary water retention capacity. Problems have not just been caused by the presence of too much water, but by too little as well: in the dry summer of 2003 the combination of low river levels and penetration of salt water far inland caused a serious water shortage.

The goal of the Biesbosch case study is to reveal the longer term effects of climate change on the Biesbosch and surrounding area and how the effects of climate change can be offset by changes in the landscape structure and innovative land management. By looking at the region as a whole, the project can demonstrate how spatial aspects relating to water throughout the whole region can be used to address the shortage of space for recreation, housing, agriculture and nature conservation – issues on which climate change is casting a new light. Specific questions include:

- How can 'depoldering' and 'polder conversion' create more room for water in the low-lying parts of the Netherlands? By giving water more space in the thinly populated parts of the 'Low Netherlands', we can not only alleviate the problems of other more densely populated areas but also make use of the energy potential in the water (natural land reclamation and peat growth).
- How and to what degree is it possible to generate the energy needs of the Biesbosch–Haringvliet region from biomass? A robust natural system in the green/blue axis offers opportunities and potential for producing biomass such as willow and reed and for extensive livestock farming.

The Biesbosch case study will produce a strategy and scenarios for a climate-proof green/blue axis from Biesbosch to Haringvliet. Concrete projects that put the strategy into practice will be drawn up for several strategic points and implemented by various coalitions of stakeholders, such as water companies, the city of Dordrecht, energy companies, nature conservation organisations and Rijkswaterstaat (the government department for public works and water management).

Kampen/IJsseldelta

The Kampen/IJsseldelta study broadly covers an area to the southwest of Kampen. The proposal is to build several thousand new houses on the southwestern edge of the town. In addition, the planned Hanzelijn railway connection is due to be built and become operational in 2012, and the capacity of the N50 trunk road will be increased. In the future, higher discharges are expected in the river IJssel, which at Kampen becomes a bottleneck that prevents the full discharge of the water in the river. Over the short term (to 2015), deepening the summer bed will be sufficient to cope with the rising discharges. But soon after that at a discharge of about 16,600 m^3/s at Lobith, where the Rhine enters the Netherlands, a bypass will be necessary. It is known that a bypass is the most sustainable option for tackling the problem of high river levels, with a major water-lowering effect (about 60 cm) extending far upstream.

The Overijssel provincial council has prepared a masterplan with the cooperation of all the stakeholders in the region and the central government partners. The plan replaces the deepening of the summer bed with the construction of a new river arm (bypass) combined with several other partly autonomous developments in the area, including house building, the construction of the Hanzelijn railway line, natural habitat restoration and recreational facilities. This alternative is more expensive than dredging the channel in the summer bed of the river, but delivers better spatial quality. Moreover, it lays a foundation for development in the longer term (from 2015), when even higher discharges are expected. The aim of the Kampen/IJssel project (integrated area development) is to bring forward the construction of a new river arm.

Climate change effects are mainly felt through the higher river discharges in the IJssel. In the long term this will be joined by a further factor as rising sea levels force up water levels in the IJsselmeer. Under prevailing westerly winds (and storms) the water in the IJsselmeer can be forced up through Ketelmeer into the IJssel. In the longer term, the existing sea defences along the lower reaches of the delta may not be high enough to withstand the water. The case study is an integrated assessment of the summer bed deepening, bypass and masterplan alternatives for Kampen/IJsseldelta.

Tilburg

Tilburg lies on a more elevated part of the Netherlands (sandy soils) and has a population of more than 200,000. The city lies in the fine network of streams and tributaries of the river Dommel, which flows into the country from the higher sandy areas to the south. The surrounding natural habitats consist largely of forest and bog pools. The region is known for its many tourist and leisure attractions, such as the Efteling theme park and the Beekse Bergen safari and fun park. Table 8 summarises the climate change issues affecting Tilburg.

The first phase of the case study, to be carried out under the Climate *changes* Spatial Planning programme, focuses on the following questions:

- What is going to change in and for our region? (content)
- How are we going to deal with these changes? And with whom? (process)

In the second phase the ideas and concepts arising in the first phase will be put into practice:

Table 8 Climate change issues affecting Tilburg

Primary effect	Secondary effect	Financial consequences	Relevance to Tilburg region
More and/or more intensive precipitation	Sewer overload	Considerable damage	Old districts in lower-lying areas in particular
More precipitation, greater water inflow	Water courses overloaded	Damage to agricultural land and natural habitat	Will affect stream systems south of Tilburg
Temporary hot weather	Health problems among vulnerable groups Appropriate measures at major events Comfort or 'H&S effect'	Specific attention to high-risk groups 'Heat breaks', 'tropical working hours', cooling costs	Hot weather plan for: • Events • Social map • Community health services Need for energy, water, greenspace and shade: climate-proof buildings
Average temperature rise	Longer agricultural growing season Longer tourist/recreation season	Boost for new crops Economic stimulus to recreation sector	Promotion of Tilburg region Economic vitality Employment

Source: de Pater and van Drunen (2006)

- Creating a vision on the spatial development of the region to 2050; adaptation and mitigation are both important and will be looked at together
- Developing a climate and opportunity plan
- Designing new climate-proof housing areas (city centre and extension)
- Restructuring old districts

The alliance of municipal councils, housing corporations, developers, water boards and resident groups is at the heart of the last two topics. Important aspects are water management, sewerage, heat-resistant buildings and the ratio between built-up areas and the open countryside. The idea of a climate alliance will also be central to the study.

What Next?

In the second phase of ARK, which will run for a year, agreements will be made with various parties on implementing the national adaptation agenda. Preparations will also be made for the third phase, which will run for a further 6 years. Under the ARK strategy, national government does not work alone, but works with local authorities, market players and other stakeholders. The aim is to put climate change into the mainstream of policy and investment decision-making, public behaviour and research programmes during the course of the third phase, making it as much a part of decision-making as financial analysis. Box 4 reviews the principles for the implementation of an adaptation agenda. The implementation agenda itself is of course specific to each location.

Examples of activities to be taken over the short term:

- Central government will take the effects of climate change into account when drawing up strategic national plans and implementing existing plans
- Central government will review the economic and social opportunities and constraints associated with shifting the weight of investment from lower-lying areas to higher sites
- Central government will draw up a list of criteria for preparing new building and infrastructure plans which appraise the effects of temperature rises, high water and flooding, wind direction and aspect
- The Ministry of the Interior will be responsible for preparing disaster emergency plans that factor in weather conditions and will calculate what they contribute to reducing the scale of damage and saving lives
- The Association of Provincial Authorities (IPO) will draw up a list of case studies to serve as demonstration projects for locations where climate change has to be taken into account
- IPO will prepare a resource guide for provincial councils to help them draw up spatial visions that incorporate the effects of climate change
- The provinces will take the effects of climate change into account when revising a regional plan or an integrated spatial and environmental plan

It will be up to the new government and the new provincial council to make progress with the integration of the effects of climate change into spatial policies and with putting them into practice. Implementation must be targeted on innovation and climate-proofing. The role of the National Programme on Adapting Spatial Planning to Climate Change is to stimulate advice and to monitor.

Box 4 Principles for adaptation measures
- Deal with the inevitable negative consequences
- Short-term local benefits
- Prevent damage
- Exploit opportunities
- Make systems more robust and/or flexible
- Start no-regret measures in good time
- Deal with uncertainties: risk management
- Government steering vs autonomous adaptation
- Increase climate awareness and climate communication
- Review options

Acknowledgements The Routeplanner research has been made possible through the contributions of the BSIK programme's Climate *changes* Spatial planning, Living with Water and Habiforum. This chapter has been reviewed by Jaap Kwadijk, Jeroen Veraart, Ekko van Ierland and Ms. Florrie de Pater.

References

Australian Greenhouse Office, in the Department of the Environment and Heritage (2006) Climate Change Impacts & Risk Management, A Guide for Business and Government, Canberra

Danish Ministry of the Environment (2005) Denmark's Fourth National Communication on Climate Change, Kɸbenhavn

De Bruijn KM (2005) Resilience and flood risk management.WL Delft Hydraulics Select series 6, Delft

European Environment Agency (EEA) (2005) Vulnerability and adaptation to climate change in Europe, Copenhagen

Gouldby B et al (2005) Language of risk. FLOODsite report T32-04-01

Ierland EC, van K, de Bruin RB, Dellink, Ruijs A (eds) (2006) A qualitative assessment of climate adaptation options and some estimates of adaptation costs, Routeplanner subprojects 3, 4 and 5, Wageningen UR

IPCC (2001) IPCC Third assessment report: Climate Change 2001. Intergovernmental Panel on Climate Change (IPCC). Cambridge University Press, Cambridge

KNMI, Royal Netherlands Meteorological Institute (2006) KNMI Climate Change Scenarios 2006 for The Netherlands. Report 2006-01, De Bilt

Kwadijk J, Klijn F, van Drunen MA, de Groot D, Teisman G, Opdam P, Asselman N (2006) Klimaatbestendigheid van Nederland: nulmeting, Routeplanner deelproject 1, WL|Delft Hydraulics

Ministry of Agriculture and Forestry Finland (2006) Finland's national adaptation strategy, Helsinki

Osinga J, Kabat P (2006) In: Veraart JA, Opdam P, Nijburg en C, Makaske B (eds) 2006 Quickscan Kennisaanbod en -leemten in Klimaatbestendigheid. Effecten, adaptatiestrategieën en maatschappelijke inbedding. Deelrapport uit het Routeplanner traject 2010–2050 in kader van de BSIK programma's Klimaat voor Ruimte, Leven met Water, Habiforum/Vernieuwend Ruimtegebruik en Ruimte voor Geo-Informatie.

Pater F. de, van Drunen MA (2006) Casestudies en Hotspots, Routeplanner deelproject 6, Klimaatcentrum Vrije Universiteit, Amsterdam

Stern N (2006) Review report on the economics of climate change. HM Treasury, UK, Available at http://www.hm-treasury.gov.uk/stern_review_report.htm

UK – Secretary of State for the Environment, Food and Rural Affairs (2006) Climate Change The UK Programme, Norwich, CM6764, SE/2006/43

U.S. Department of State (2002) US Climate Action Report 2002, Washington, DC

van Ierland EC, Bruin Kd, Dellink RB, Ruijs A (2007) A qualitative assessment of climate adaptation options and some estimates of adaptation costs. Wageningen Universiteit en Research, Wageningen

Veraart J, Opdam P, Nijburg C, Makaske B, Brinkman S, Pater F de, Luttik J, Meerkerk J, Leenaers H, Graveland J, Wolsink M, Klijn EH, Neuvel J, Rietveld P (2006) Quickscan, Kennisaanbod en -leemten in Klimaatbestendigheid, Effecten, adaptatiestrategieën en maatschappelijke inbedding, Routeplanner deelprojecten 2, Wageningen UR

WLO (2006) Welvaart en Leefomgeving – een scenario'studie voor Nederland in 2040, concept 13 Apr, CPB, MNP en RPB, 135 pp

The Strategic Role of Water in Alleviating the Human Tragedy Associated with HIV/AIDS in Africa

Jeanette Rascher, Peter Ashton and Anthony Turton

Introduction

The continuing HIV/AIDS pandemic across sub-Saharan Africa has had an enormous impact on societies, and the number of people dying from AIDS-related diseases has reached truly alarming levels in several countries. The series of adverse impacts appears likely to continue – especially in southern Africa – where HIV prevalence rates have reached heights seen nowhere else in the world, and little evidence is available to suggest that national prevalence rates have declined in real terms (Van Dyk, 2001; Walker et al. 2004; Marais, 2005; UNAIDS, 2006a). The situation has been worsened in recent months by the emergence of a drug-resistant strain of tuberculosis, one of the opportunistic diseases of HIV/AIDS, which has now placed an increased health burden on people in southern Africa.

The HIV/AIDS pandemic threatens the future of African societies while the accompanying disintegration of social systems contributes to a social and ideological environment that favours the spread of the HIV virus. The most vulnerable victims of HIV/AIDS comprise the potentially most economically productive age group of the population, many of whom are living in extreme poverty (Whiteside and Sunter, 2000; Ramasar and Erskine, 2002; Kamminga and Schuringa, 2005; Marais, 2005).

Importantly, the HIV/AIDS pandemic is not simply a health issue that calls for commitment from governments to provide assistance in the form of antiretroviral treatment to infected individuals. Instead, it is now widely acknowledged that HIV/AIDS is a development problem that affects the whole fabric, structure and future of African societies (Ashton and Ramasar, 2002; Barnett and Whiteside, 2002; Ramasar and Erskine, 2002; Whiteside and Sunter, 2000; Kamminga and Schuringa, 2005), and there is a high and growing probability that massive political, ecological and social changes will occur during the next few years (Friedman et al. 2006). Efforts directed at minimising the destructive social crises

---

J. Rascher (✉)
Council for Scientific and Industrial Research (CSIR) –
Natural Resources and the Environment, Water Resource Governance Systems,
PO Box 395, Pretoria, 001, South Africa
e-mail: jrascher@csir.co.za

and maximising the potentially constructive outcomes of the pandemic have recognised the indispensable role that the provision of wholesome supplies of water and appropriate sanitation services has in strengthening the ability of communities to withstand the impacts of increased AIDS-related mortality and helping to ensure a longer life for people living with HIV/AIDS (Ashton and Ramasar, 2002; Van Wijk, 2003; Kamminga and Schuringa, 2005; UN-HABITAT, 2006).

In order to understand more clearly the intricate linkages between HIV/AIDS and water, this chapter reviews the demographic impacts of HIV/AIDS in Africa, examines the socio-economic impacts of the disease related to water and evaluates the strategic role of water to provide hope for those that are infected by HIV/AIDS.

Impacts of the HIV/AIDS Pandemic

Demographic Impacts

From the data in Table 1 it is evident that the HIV epidemic is worsening at an alarming pace. According to statistics it is estimated that the number of adults and children in sub-Saharan Africa living with HIV has, since 2004, increased by 1.1 million and that the number of adults and children newly infected with HIV has increased by 0.2 million. Statistics also reveal that the annual number of deaths due to AIDS between 1997 and 2004 was 0.2 million. Similarly, statistics for North Africa and the Middle East show an increase. Between 2004 and 2006, adults and children living with HIV have increased from 400,000 to 460,000, adults and children newly infected with HIV have increased from 59,000 to 68,000, while adult and child deaths due to AIDS have increased from 33,000 to 36,000 (UNAIDS, 2006a,b).

'Of what use are statistics if we do not know what to make of them?' was the question asked by Florence Nightingale during the Crimean War (Marais, 2005). Today, more than 150 years later, the same question can justifiably be asked about the accumulation of HIV/AIDS statistics. Since the start of the HIV/AIDS pandemic in the early 1980s, the health of African populations has become increasingly

Table 1 HIV/AIDS statistics and features as at end of 2006 (Source: UNAIDS, 2006a)

Region	Year	Adults and children living with HIV	Adults and children newly infected with HIV	Adult prevalence %	Adult and child deaths due to AIDS
Sub-Saharan Africa	2006	24.7 million	2.8 million	5.9%	2.1 million
	2004	23.6 million	2.6 million	6.0%	1.9 million
North Africa and Middle East	2006	460,000	68,000	0.2%	36,000
	2004	400,000	59,000	0.2%	33,000

adversely affected. This is particularly visible in increased rates of sickness and death at ages where normal rates of morbidity and mortality are low, mainly within the economically active and productive age groups of a population (Hooper, 2000; Janse van Rensburg, 2000; Whiteside and Sunter, 2000; Ashton and Ramasar, 2002; Barnett and Whiteside, 2002). Initially, very little effort was directed towards utilising these data to fully understand the scale, trends and trajectories of the HIV/AIDS pandemic. Many of the earlier warnings were ignored, disputed or dismissed (Ashton and Ramasar, 2002; Marais, 2005) and a lack of decisive and concerted action enabled the pandemic to exert devastating impacts on the social and economic fabric of most countries in sub-Saharan Africa.

In those cases where epidemiological statistics did receive attention, epidemiologists often used predictive models that were derived from other disease epidemics to predict the spread of HIV/AIDS. These early models were adapted to fit the reported patterns of HIV/AIDS cases and it was concluded that the numbers of new HIV/AIDS cases would soon decline rapidly. When this did not happen, it was realised that curve fitting models did not work as well for HIV/AIDS because the transmission dynamics of HIV are much more complex than other diseases. Moreover, HIV/AIDS has a distinctive ability to continue moving from one group to another within a geographic region as a result of population dispersal and the heterogeneous contact patterns that spread infection (Whiteside and Sunter, 2000). No one predicted that the pandemic would worsen at such an alarming rate (Ashton and Ramasar, 2002; Barnett and Whiteside, 2002; Ramasar and Erskine, 2002), sometimes increasing by as much as tenfold in 5 years as has been recorded in several southern African countries (Sufian, 2000). Today, more than 20 years since the start of the pandemic, the collection of reliable data is still compromised in many African countries by political reasons, poor governance systems, fear and stigmatisation, under-reporting, a lack of health facilities and the difficulty of collecting accurate data in remote rural areas where many people live. Consequently, organisations such as UNAIDS have to generate estimates that are based on relatively low levels of data, many of which are still unreliable, which in turn affects the accuracy of population projections. In addition, social epidemiologic analyses of the HIV/AIDS pandemic have also been neglected and very little attention has been paid to the role of social change and its potential importance in prevention efforts (Myer et al. 2004). In effect, therefore, the pandemic has been 'allowed' silently to take its toll, not only in terms of the high morbidity and mortality rates, but also in terms of the enormous non-linear impacts that affect every single aspect of society (Box 1).

Box 1 Influence on statistics

Stigmatisation and discrimination influence statistics
The social stigma and discrimination related to HIV-positive victims remain an enormous barrier to the fight against AIDS. Fear of discrimination often prevents people from being tested, seeking treatment and publicly admitting

> their HIV status. The fear and prejudice that lies at the core of the HIV/AIDS discrimination needs to be tackled at both community and national levels.
> Source: http://www.avert.org/aidssouthafrica.htm.

The HIV/AIDS pandemic is a global problem and is not limited to the developing world. However, the overwhelming majority of HIV-positive people, some 95% of the global total, live in the developing world, where poverty, inadequate health and education systems, gender inequality and limited resources contribute to worsening the adverse impacts of the virus (Whiteside and Sunter, 2000; UNAIDS, 2006a). The burden of communicable diseases such as AIDS, tuberculosis and malaria is much greater in the poorer parts of the world where the lower life expectancy of people is a partial reflection of their exposure to infectious diseases (Barnett and Whiteside, 2002). The demographic impact of HIV/AIDS in Africa needs to be evaluated (Fig. 1) against the background of the global situation to fully appreciate the devastating situation, especially in sub-Saharan Africa.

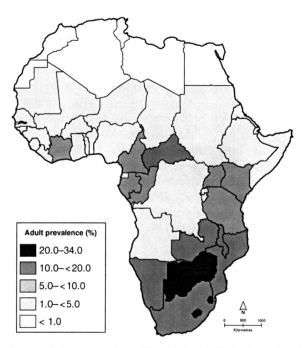

Fig. 1 Adult HIV prevalence (%) in Africa, 2005 (Redrawn from UNAIDS, 2006a)

The Global Situation

A global view of HIV/AIDS prevalence statistics reveals that approximately 38.6 million people were living with HIV in 2005 (UNAIDS, 2006a). The number of people living with HIV continues to grow and an estimated 4.3 million adults and children were newly infected with HIV in 2005; this figure is about 400,000 more than the estimate of newly infected individuals for 2004. Globally, new HIV infections are mostly concentrated among young people between the ages of 15 and 24, and overall, young people accounted for 40% of all new HIV infections in 2006 (UNAIDS, 2006a).

Sub-Saharan Africa

In sub-Saharan Africa, adult prevalence rates are growing and recent estimates indicate that almost 25 million people are living with HIV/AIDS (UNAIDS, 2006a); this figure represents 63% of the global total. According to current statistics, more than 15 million Africans have died from AIDS and AIDS-related diseases since the beginning of the pandemic. It is estimated that 2 million Africans have died from AIDS and AIDS-related diseases during 2006 (UNAIDS, 2006a). This figure represents almost three-quarters (72%) of all AIDS-related deaths globally. During 2005, an estimated 2 million adults and children died as a result of AIDS in sub-Saharan Africa (Fredricksson and Kanabus, 2006; UNAIDS, 2006a,b).

Box 2 An increase in the 'death business'

The number of AIDS deaths result in a sharp increase of the 'death business'
The number of AIDS-related deaths in Africa resulted in a sharp increase of the death business, which includes funeral services and coffin makers (source: http://www.caj.ca). In Zambia, an area on the western fringes of the central business district in the capital, Lusaka, has been dubbed 'Death Valley' in recognition of the concentration of businesses like undertakers, coffin manufacturers and funeral insurance companies. Although the capital has 6 registered funeral parlours, a further 21 unregistered parlours have sprung up as a result of the growing demand for funeral services.
Source: http://www.plusnews.org.

Southern Africa continues to bear the brunt of the HIV/AIDS pandemic as the prevalence rates continue to rise. In 2005, Swaziland had the highest adult HIV prevalence rate in the world at 33.4% (UNAIDS, 2006a). South Africa also falls into the category of 'high HIV prevalence countries' (Chetty and Michel, 2005; Phaswana-Mafuya and Peltzer, 2005; UNAIDS, 2006a), with some 5.5 million people, including 240,000 children younger than 15 years of age, living with HIV in 2005 (UNAIDS, 2006a). The latest official mortality data indicate that the total number of deaths from all causes in South Africa increased by 79% from 1997 to 2004, suggesting that South Africa has now reached the stage where increasing numbers of people are dying of AIDS-related diseases (Dorrington et al. 2001; South African Medical Research Council, 2005; UNAIDS, 2006a). The increasing death toll has driven average life expectancy below 50 years in three of South Africa's provinces, namely Eastern Cape, Free State and KwaZulu-Natal (Dorrington et al. 2006).

HIV data gathered at South African antenatal clinics continue to show a rising trend in HIV infection levels among pregnant women (UNAIDS, 2006a). As in the rest of sub-Saharan Africa, the epidemic in South Africa continues to have a disproportionately greater effect on women, with young women between the ages of 15 and 24 being four times more likely to be HIV infected than are young men of a similar age (UNAIDS, 2006a). According to the 2005 national HIV household survey in South Africa, one in three women aged between 30 and 34 were living with HIV in 2005, as were one in four men aged between 30 and 39. In addition, high HIV infection levels were also found among men older than 50 years, more than 10% of whom tested HIV positive. In Namibia, the national adult HIV prevalence was estimated at 19.6% in 2005 while the national adult HIV infection levels for Botswana and Lesotho are also high at between 20% and 24% (UNAIDS, 2006a).

Box 3 Rape and the high prevalence of HIV and social disruption

Rape plays a significant role in the high prevalence of HIV and social disruption

Many black South African men infected with AIDS erroneously believe that by having sex with a virgin, even a child or baby, they will be cured of AIDS or their HIV infection. This misconception is fuelling what is already one of the highest child sexual exploitation rates in the world. According to crime statistics for the year 2000, the latest report by South Africa's Police Service, children are the victims of 41% of all rapes and attempted rapes reported in the country. Over 15% of all reported rapes are against children aged under 11, and another 26% against children aged 12–17. The trend is worsening. Babies as young as only a few months old are being raped almost daily (source: http://www.worldnetdaily.com). In a 2006 study of 1,370 South African men, nearly one-fifth of them revealed that they had raped a woman.
Source: http://www.avert.org/aidssouthafrica.

In East Africa, the general trend of a stabilising or a declining HIV prevalence appears to be continuing and HIV infection levels happen to be lower in this region than in the south of the continent. National HIV prevalence among pregnant women has declined in Kenya and Tanzania and, to a slightly lesser extent, in Rwanda. Programmes to prevent the spreading of HIV/AIDS in Uganda have initially succeeded in lowering its high infection rates. However, the latest research results indicate a possible erosion of the gains that Uganda made against AIDS in the 1990s, while HIV infection levels among pregnant women in Burundi's capital, Bujumbura, have also shown a sudden increase in 2005 (UNAIDS, 2006a).

In Central Africa, adult HIV/AIDS prevalence data indicate that Cameroon (5%) and the Central African Republic (11%) are the most heavily affected (UNAIDS, 2006a). In Cameroon, infection levels are highest in the northwest and eastern regions (almost 9%) and lowest in the north of the country (2% or lower) (UNAIDS, 2006a). As indicated elsewhere, these numbers are regarded as underestimates because they are based on limited surveillance data and estimates derived from discussions with local governments, where many politicians are reluctant to allocate domestic resources to HIV/AIDS programmes.

In West Africa, the urban parts of Burkina Faso, Côte d'Ivoire and Ghana are showing signs of declining HIV prevalence, while adult HIV prevalence appears to be increasing in Mali (UNAIDS, 2006a).

Africa North of the Sahara

In the mainly Muslim countries of North Africa, inadequate HIV surveillance has had a distinctly adverse effect on the accuracy and quality of the HIV/AIDS data that are available. As a result, the HIV prevalence data for these countries seem very low when compared to those of southern African countries. Despite the tragedy that is unfolding in southern Africa, only a few Islamic authorities north of the Sahara seem to have responded to the alarm (Sufian, 2000; Eberstadt and Kelley, 2005; Kuppusamy, 2005) and few leaders have realised that Islamic culture and beliefs do not make a population immune to HIV/AIDS. Until recently, many Islamic authorities regarded HIV as a 'disease of the West' and considered that the region was unlikely to be subjected to the spread of the virus because Islamic rules require no sex outside of marriage (Williams, 2005). After years of neglect and denial, there are recent signs in some of the North African countries of a dawning awareness that HIV/AIDS is becoming a domestic dilemma. It is not yet clear what impact this will have as awareness means acceptance as the 'respectable' face of AIDS in North Africa. The only people that are willing to discuss the disease are men who have contracted HIV through blood transfusions or women who have been infected by their husbands. Only a few are willing to talk openly about commercial sex as it is illegal in much of the region. Denial, stigma and discrimination are still enormous problems, not least among the region's medical professionals (El Feki, 2006). The Joint United Nations Programme on HIV/AIDS (UNAIDS) estimates the total HIV-positive population of North Africa, the Middle East and predominantly Muslim Asia at nearly 1 million people (Eberstadt and Kelley, 2005).

UNAIDS data on the number of people living with HIV/AIDS are completely missing for a number of African countries. The lack of adequate data and the deliberate avoidance of HIV/AIDS discussions are hampering appropriate responses to the very real probability that HIV/AIDS transmission rates will increase dramatically once a certain threshold has been reached (Sufian, 2000; Kuppusamy, 2005). According to Joan MacNeil, senior HIV/AIDS specialist for the Global HIV/AIDS Programme of The World Bank Group, an epidemic threshold is reached 'when enough critical mass of risk behaviours and contributing biological factors exists in a population to sustain an epidemic' (Sufian, 2000; Kuppusamy, 2005). This means that an epidemic will be sustained and grow if an infected individual infects more than one additional person in their lifetime. Importantly, however, the threshold can be dramatically influenced by the size of risk populations, the type and frequency of risk behaviour and the presence of other sexually transmitted infections (Kuppusamy, 2005). If governments delay action, the trend witnessed in other regions of Africa will likely recur in North Africa (Sufian, 2000; Kuppusamy, 2005).

Box 4 Effect on HIV/AIDS governance

Denial affects HIV/AIDS governance
Many people argue that the response to HIV/AIDS in South Africa has been hampered by 'AIDS denialism', a minority scientific movement that refutes the orthodox idea that HIV causes AIDS. Some leading figures in South Africa have flirted with this school of thought, much to the dismay of AIDS activists. Former President Mbeki has consistently refused to acknowledge that HIV is the cause of AIDS; he argues that HIV is just one factor among many that might contribute to deaths resulting from immunodeficiency, alongside others such as poverty and poor nutrition.
Source: http://www.avert.org/aidssouthafrica.htm.

Socio-economic Impacts of the HIV/AIDS Pandemic Related to Water

Against the background of the demographic data on HIV/AIDS, it is important to examine the close link between these statistics and the impacts that flow from it. A clear understanding of the relationship between the HIV/AIDS demographic characteristics (such as population numbers, growth rates, structures of populations, structure by gender and age, key indicators such as birth, death and fertility rates, life expectancy and infant and child mortality, health status and health needs) is essential for determining the amount of health care and infrastructure resources required by the population (Gilbert et al., 1997; Barnett and Whiteside, 2002) and for determining the likely mitigatory actions required to alleviate the negative impacts on all other spheres of life.

The unusual levels of morbidity and mortality caused by HIV/AIDS exert continuous changes on population dynamics. Therefore, ongoing studies of human groups with reference to their size, composition and distribution are needed if we are to understand, cope with and mitigate these negative effects fully. Past oversimplification of HIV/AIDS demographic statistics has contributed to widespread ignorance about the devastating socio-economic impacts that the pandemic would have on societies in sub-Saharan Africa. In many cases, HIV/AIDS was considered to be a purely medical or health issue, without significant links to every other sector of society (Ashton and Ramasar, 2002; Marais, 2005). Today, these earlier misconceptions have been replaced by a growing awareness that HIV/AIDS has slowly but surely caused a complex number of subtle and overt changes in society that have resulted over time in a series of accumulated impacts that, collectively, represent the single biggest barrier to development in sub-Saharan African countries (Barnett and Whiteside, 2002). Inadequate or inappropriate forms of development increase people's susceptibility to HIV transmission and their vulnerability to the impact of AIDS (Ashton and Ramasar, 2002; Barnett and Whiteside, 2002; Ramasar and Erskine, 2002).

The most direct linkage between health and general development can be found in the strategic roles that supplies of clean water and appropriate sanitation systems play in alleviating many of the adverse effects of the HIV/AIDS pandemic. A lack of access to clean, safe water is the leading cause of death and disease in developing countries, especially individuals with compromised immune systems. For millions of people in Africa, access to sufficient supplies of clean water is a matter of life and death; the situation is greatly aggravated by the impacts of HIV/AIDS and poverty, and the brunt of these combined impacts is borne by individual households (Ashton and Ramasar, 2002; Ramasar and Erskine, 2002; Van Wijk, 2003).

Sectoral Impacts

The HIV/AIDS epidemic has placed increased pressure on the health sector in different countries. In particular, as HIV prevalence rates rise, the demand for care for those living with HIV also rises, as does the toll among health workers. The challenge of sustaining and extending services in developing countries where the number of skilled people is often limited puts sectors, and especially the health sector, under considerable pressure (Van Wijk, 2003; Kamminga and Schuringa, 2005; IRC, 2006).

Recent research has shown that, on average, HIV-positive patients in sub-Saharan Africa stay in hospital four times longer than other patients. In South Africa, for example, it is estimated that patients affected by HIV and AIDS account for almost 60–70% of total hospital expenditure (Fredericksson and Kanabus, 2006). In poorer African countries, hospitals are struggling to cope as there are often not enough beds available. The situation is complicated by the fact that large numbers of healthcare professionals have also been affected by the pandemic. Botswana, for example, lost 17% of its healthcare workforce due to AIDS between 1999 and 2005, while a study conducted in one region of Zambia found that 40% of midwives were HIV positive

(UNAIDS, 2006a). The lack of professional healthcare capacity therefore places pressure on home-based care for the terminally ill, especially to those living in rural areas where there is inadequate water and sanitation.

The water and sanitation sector has also been seriously affected by the HIV/AIDS pandemic because demands for these services bring issues of access and affordability to the fore (Ashton and Ramasar, 2002; Kgalushi et al. 2003). Approximately 300 million people in Africa lack access to adequate water supplies, and an estimated 313 million people lack access to appropriate sanitation. In several of the poorest countries, such as Ethiopia, Mozambique, Tanzania, Uganda and Zambia, a low coverage of safe water supply and sanitation is accompanied by a high HIV/AIDS prevalence (Kamminga and Schuringa, 2005). According to the UN Millennium Project, more than 50% of Africans suffer from water-related diseases such as cholera and infant diarrhoea. It has been estimated that the number of people without adequate water supply and access to sanitation could double by the 2015 MDG target if the 'business as usual' approach continues (UNSGAB, 2006).

Impacts on Households

The worst impacts of HIV/AIDS in communities are felt at the household level since this is where direct costs are borne and mitigation interventions must be located (Obi et al. 2006). Poor households are extremely vulnerable to the HIV pandemic and also tend to suffer the worst consequences because they are seldom well equipped to mitigate adverse impacts. The traditional 'safety net' of poor households becomes greatly weakened when income earners die or are forced to stay at home to care for sick relatives. Many of the victims that are dying have surviving partners who are infected and who are also in need of care. In such cases, desperate relatives that lack access to appropriate information and facilities often spend their savings on inappropriate cures or treatments, worsening the burden of poverty on already poor households (Van Wijk, 2003; IRC, 2006; Obi et al. 2006). A recent study in South Africa found that households that had experienced illness or death in the recent past were more than twice as likely to be poor than non-affected households and were more likely to experience long-term poverty (UNAIDS, 2003).

In many cases, the presence of HIV/AIDS within a family unit means that the household structure will dissolve, as parents die and children are sent to relatives for care and upbringing (Obi et al., 2006). When no adult family members are left to care for orphaned children, the eldest children have to head households and care for their younger siblings. Once orphaned, these children are grieving and struggling to survive without proper parental care (Barnett and Whiteside, 2002) and are then likely to face worsening poverty and poorer health. In communities that have been hard-hit by HIV/AIDS, households not directly incurring a death may nevertheless be affected by having to take in orphans, losing access to resources owned by kin-related 'afflicted' households or by transferring scarce resources to afflicted households (Jayne et al., 2006). The situation is worsened by the fact that many of the orphans may also be HIV positive (Campbell et al. 2005). At the end of 2005, around 2 million children in sub-Saharan Africa were living with HIV, representing more than 85% of all children living with HIV worldwide. The vast majority of these children will have become infected with HIV during pregnancy or by being breast-fed by HIV-positive mothers (UNAIDS, 2006a).

> **Box 5** Unintended consequence of HIV/AIDS mitigation strategy
>
> *Effect of nitrate*
> One of the unintended consequences of HIV/AIDS mitigation strategies can be found in the nitrate-rich groundwater areas of South Africa. HIV/AIDS-positive women who breast-feed their infants are sometimes advised to rather bottle-feed with formula. While this prevents mother-to-child transmission of HIV, it inadvertently exposes infants to elevated levels of nitrate in groundwater, which can lead to a condition known as methaemoglobinaemia.
> Source: Colvin and Genthe (1999)

Growing reliance on home-based care is escalating and, while this places increased pressure on women, it also entrenches gender inequalities. Women in sub-Saharan Africa that are normally the caregivers of their families now also have to take care of HIV/AIDS patients while sometimes being HIV positive themselves (Campbell et al. 2005). At the same time they are responsible for cultivating crops, provisioning of cash income, cleaning and cooking – often without access to clean water and proper sanitation. Because of the additional work involved in caring for the sick, girls are taken out of school to take on a wider range of household and domestic responsibilities. School enrolment rates for girls therefore tend to decrease in communities with high HIV prevalence rates. This, in turn, decreases access to education, information, knowledge and income-generating opportunities, which can increase susceptibility to infection.

Caregiving has also heavily impacted older women emotionally, physically and financially as adult children frequently return home when they become sick with HIV/AIDS. In most African cultures, older women are normally helped and supported by their adult children, but HIV/AIDS has placed a heavy burden on them as they now have to care first for their dying children and sometimes grandchildren and then their orphaned grandchildren, which causes physical and emotional exhaustion that worsens their own medical problems. They therefore often have to rely on children to assist with caregiving and household chores.

The management of health and the provisioning of home-based care for people living with HIV/AIDS not only presents a socio-economic and emotional challenge to those that are living in poverty, but is worsened by the lack of safe and sufficient water and sanitation that is indispensable for people living with HIV/AIDS (Potter, 2006).

The burden of care is made heavier by the inaccessibility of water close to the family dwelling. Numerous poor households also depend on small-scale agriculture for their food and livelihood, but family members may be too sick, too young or too poor to farm or feed themselves. The situation is worsened when water supplies fail as food gardens die and livestock suffer which results in more opportunities to spread the virus.

Often water has to be carried long distances to the house which takes time and effort, a burden borne mainly by women. Time spent on fetching water reduces the time available to care for the sick and in many instances family members that are caregivers or patients that are too ill or too weak are obliged to buy water from vendors at a higher cost. A heavy load is also placed on children that have to fetch water when they return from school and often having to walk long distances, carrying heavy containers with water.

Providing for daily water needs is a burden on households with inadequate services in a number of other ways, in addition to the direct health threats. Most HIV-infected people living in poverty cannot afford to buy clean drinking water. Female family members and caregivers therefore mostly struggle to address the infected person's physical needs such as safe drinking water for taking medicine, bathing the patient, washing their soiled clothes and linen and keeping the environment clean. They often have no other choice but to make use of water that does not meet basic health standards by fetching untreated water from streams, dams and other unreliable water sources. This poses a high risk to all household members, including patients who are already immuno-compromised by HIV/AIDS. Illnesses are usually contracted through contaminated water which causes diarrhoea, nausea, vomiting, and weight loss. In the case of infants, the situation can be detrimental as the chance of a child dying from diarrhoea rises when formula feeds are not prepared with clean water or when cleaning and water handling practices are not hygienic. In urban and urban fringe areas water is often only available from vendors at a price that is usually several times more expensive than the water provided through formal services and of poor quality (Van Wijk, 2003; Kamminga and Schuringa, 2005; IRC, 2006; Phaswana-Mafuya, 2006) and mostly unaffordable for the poor.

In addition to inadequate access to clean water, many rural areas in Africa lack access to adequate sanitation and experience a wide range of health problems such as typhoid, bilharzia, malaria, cholera, worms, eye infections and skin diseases. The situation is worse for immuno-compromised individuals who are now more vulnerable to infections and illnesses that they would normally be able to resist for a longer period of time (Tladi et al. 2002; Phaswana-Mafuya, 2006).

HIV/AIDS also exerts psychological and emotional impacts on individuals and households where stigma and discrimination are commonly manifested in the form of blame and punishment. Infected individuals often experience stigma and discrimination in the home, and women are often more likely to be badly treated than men or children (Bharat and Aggleton, 1999). Families may reject people living with HIV/AIDS, not only because of their HIV status, but also because HIV/AIDS is associated with promiscuity, homosexuality, and drug use (Mujeeb, 1999). In many cases, HIV/AIDS-related stigma and discrimination has been extended to families, neighbours and the friends of people living with HIV/AIDS. This 'secondary' stigmatisation and discrimination has played an important role in the social isolation of those affected by the epidemic, such as the children and partners of people living with HIV/AIDS. In a community context, people often shun or gossip about those perceived to have HIV/AIDS and tend to exclude those that are infected from community-based decision-making. In more extreme cases, it has taken the form of violence (Nardi and Bolton, 1991).

Against the background of the socio-economic impacts on individuals, households and communities, it is understandable that the burden of the disease, combined with HIV/AIDS, increasing poverty, inequity and a lack of water and sanitation, increasingly contribute to a 'hopeless situation', especially for women that have to bear the double brunt of being HIV positive and at the same time be responsible for caregiving to other family members that are infected (Lundberg et al. 2002; Baylies, 2002).

The Strategic Role of Water in Providing Hope for Those Living with HIV/AIDS

Although water is not a cure for HIV/AIDS, it is the most powerful healing substance known to man which can contribute to slowing down the progress of the disease, allowing those that are infected to live longer, to live more comfortably, and to be productive for a longer period of time. The world's responsibility is therefore to ensure that the poor are provided with clean water to give them hope for life as 'we shall not finally defeat AIDS, tuberculosis, malaria, or any of the other infectious diseases that plague the developing world until we have also won the battle for safe drinking water, sanitation and basic health care' (Kofi Annan, United Nations Secretary General confirmed this statement in 2005).

The message of hope for those that are infected primarily lies in acknowledging that appropriate access to supplies of clean water is not only a fundamental human need but is also a human right (Gleick, 2004). At the same time, access to water enables individuals and communities to access additional health and economic benefits (Obi et al., 2006). It has also long been understood that health is a key determinant of development, 'just as development is undoubtedly the best medicine for an unhealthy nation' (Gilbert et al. 1997; Barnett and Whiteside, 2002; Van Wijk, 2003).

The human tragedy and devastation associated with the HIV/AIDS pandemic and the associated problems caused by a lack of access to clean water and appropriate sanitation systems – as well as poverty – can no longer be ignored by global and national action plans or by any organisation with an interest in Africa's development and the Millennium Development Goals. The United Nations Millennium Declaration confirmed the central role of water and sanitation in sustainable development, emphasising the major contribution that access to safe drinking water and adequate sanitation can make to poverty alleviation and subsequent disease reduction and the central importance placed on the adoption of national policies and strategies for integrated water resources management (Van Wijk, 2003; DFID, 2006; UNDP, 2006; UN-HABITAT, 2006; World Bank, 2006).

Tackling the HIV/AIDS, water and poverty crisis in Africa is a long-term task that requires sustained effort and planning, both among the international community and within African countries themselves.

It is important that the water and sanitation sector ensures that an HIV/AIDS-sensitive institutional framework is given the highest priority. Strategies should be developed for the integration of HIV/AIDS awareness into all their projects, programmes and activities (Kamminga and Schuringa, 2005; DFID, 2006; IRC, 2006; UNAIDS, 2006b; UN-HABITAT, 2006) and for the reduction of discrimination and stigmatisation of the disease to reduce the violation of human rights (Van Wijk, 2003; Kamminga and Schuringa, 2005). Awareness should be raised among national policymakers, as well as among employees of the water and sanitation service sectors, on how to address HIV/AIDS through these water and sanitation initiatives (UN-HABITAT, 2006).

The fight against HIV/AIDS calls for a multi-sectoral approach 'in which the water and sanitation sector finds its place and plays its role effectively' (UN-HABITAT, 2006) to ensure that demographic and socio-economic changes forced by HIV/AIDS are taken into account (Kamminga and Schuringa, 2005; DFID, 2006; IRC, 2006; UNAIDS, 2006b). The water and sanitation sector should form strategic partnerships with specialised HIV/AIDS programmes and interventions and support HIV/AIDS initiatives focusing on the most vulnerable segments of communities, such as women, children and the elderly (UN-HABITAT, 2006).

In particular, attention needs to be paid to ensure that poorer communities, who experience difficulty in paying for service delivery, receive assured water supplies (UN-HABITAT, 2006) as 'we must begin by recognising that welfare is a global common good' for all (Barnett and Whiteside, 2002) and that it is therefore necessary for governments to ensure that basic human dignity is restored (IRC, 2006).

It is now widely accepted in the water and sanitation sectors that the availability of safe water and sanitation does not automatically lead to improvements in health but that hygiene promotion and appropriate hygienic behaviours are required as well (Van Wijk, 2003; Phaswana-Mafuya, 2006). National efforts to expand public awareness of the dangers associated with untreated water must therefore also include hygiene awareness and education (UN-HABITAT, 2006). Interventions need to make full use of the opportunities provided in the school curriculum for integrated health and life skills education and to link this with information around good sanitation, water care and HIV/AIDS. Hygiene education must also be specifically targeted at caregivers and volunteers involved in home-based care (Kamminga and Schuringa, 2005; Phaswana-Mafuya, 2006).

The testing and implementation of reliable water treatment processes that do not require supervision or management interventions should be given high priority. It will help to reduce the potential health risk associated with ineffective water treatment that can be expected as a result of increased mortality of operators of water treatment works (Ashton and Ramasar, 2002; UN-HABITAT, 2006).

Community-based organisations that are dedicated to social equity, provision of essential social services such as water supply and community health promotion that are the main pillars of HIV competence (Kamminga and Schuringa, 2005) need to be strongly promoted. The role of community-based structures should also be re-identified in the new context of municipal service delivery. Equally, local structures that promote accountability by service providers can play a decisive role in mitigating some of the adverse impacts of HIV/AIDS as there is an immense reservoir of goodwill, experience and positive thinking that must be mobilised if the needs of the poor are to be addressed effectively (Kgalushi et al. 2003).

Ongoing training of additional staff in the water and sanitation sector is necessary to ensure that the same level of skilled staff such as operators at water treatment works and sewage treatment works are available to avoid the accompanying deterioration in the quality of potable water supplies in urban and rural areas when employees are lost as a result of AIDS deaths. People living with HIV/AIDS should also be employed effectively in water and sanitation improvement programmes where they can act as peer educators, helping to break down prejudices and providing additional income-generation opportunities (Van Wijk, 2003; Kamminga and Schuringa, 2005).

The situation in Africa will only start to improve when serious global and national steps are taken to address the water and sanitation situation in the context of HIV/AIDS and the increasing poverty related to the pandemic. Concerted development and implementation of the appropriate strategies and cooperation between global institutions, national governments, communities and individuals is crucial for achieving the Millennium Development Goals that have been set for 2015.

Conclusion

The ever-increasing HIV/AIDS epidemic is causing significant losses of human capital that pose a serious threat to all forms of development in Southern Africa. Efforts directed at minimising the destructive social crises and maximising the potentially constructive outcomes of the pandemic have recognised the indispensable role that the provision of wholesome supplies of water and appropriate sanitation services have in strengthening the ability of communities to withstand the impacts of increased AIDS-related mortality and helping to ensure a longer life for people living with HIV/AIDS. Tackling the HIV/AIDS, water and poverty crises in Africa is a long-term task that will require sustained effort and planning, both among the international community and within African countries themselves. It is, however, alarming to note that the connection between HIV/AIDS and water resource management has not yet found its rightful place in science literature on the subject. As the tragedy of the pandemic unfolds, we are constantly being confronted by challenges to our prevailing paradigms. If we as water professionals are to play a meaningful role in attenuating or reducing the impact of the pandemic, we will all need to broaden our perspectives on HIV/AIDS and collaborate more closely to ensure that our knowledge and insights are fully deployed to combat the HIV/AIDS pandemic. If we fail to achieve this high level of professional synergy and cooperation, the tragic consequences will continue to haunt the continent.

References

Ashton PJ, Ramasar V (2002) Water and HIV/AIDS: Some strategic considerations in Southern Africa. In: Turton AR, Henwood R (eds) Hydropolitics in the developing world: A South African perspective, African Water Issues Research Unit, Pretoria, p. 217, 219

Barnett T, Whiteside A (2002) Aids in the twenty-first century: Disease and globalization, Palgrave MacMillan, New York, p. 4, 20, 67, 160, 167, 193

Baylies C (2002) The impact of AIDS on rural households in Africa: A shock like any other? Dev Change 33(4):611–632

Bharat S, Aggleton P (1999) Facing the challenge: Household responses to AIDS in India. AIDS Care 11:33–46

Campbell C, Nair Y, Maimane S, Sibiya Z (2005) Home-based carers: A vital resource for effective ARV roll-out in rural communities? AIDS Bull (Mar 2005) 14(1):22–27. Available (online) at website: http://www.mrc.ac.za/aids/march2005/homebased.htm

Chetty D, Michel B (2005) Turning the tide: A strategic response to HIV and AIDS in South African Higher Education: HEAIDS Programme Report 2002–2004. Pretoria: SAUVCA

Colvin C, Genthe B (1999) Increased Risk of Methaemoglobinaemia as a Result of Bottle Feeding by HIV Positive Mothers in South Africa. Proceedings of the International Association of Hydrogeologists conference, Melbourne

DFID (British Department of International Development) (2006) Available (online) at website: www.dfid.gov.uk

Dorrington R, Bourne D, Bradshaw D, Laubscher R, Timaeus IM (2001) The impact of HIV/AIDS on adult mortality in South Africa. South African Medical Research Council Technical Report, Johannesburg

Dorrington RE, Johnson LF, Bradshaw D, Daniel TJ (2006) The Demographic Impact of HIV/AIDS in South Africa. National and Provincial Indicators for 2006. Centre for Actuarial Research, South African Medical Research Council and Actuarial Society of South Africa, Cape Town. Available (online) at website: www.commerce.uct.ac.za.care

Eberstadt N, Kelley LM (2005) The muslim face of aids. Available (online) at website: http://www.frontpagemag.com

El Feki S (2006) Middle eastern aids efforts are starting to tackle taboos. Lancet 367(9155):975–976

Fredricksson J, Kanabus A (2006) The impact of HIV/AIDS on Africa. In: "Avert" Available (online) at website: http://www.avert.org/aidsimpact.htm

Friedman SR, Kippax SC, Phaswana-Mafuya N, Rossi D, Newman CE (2006) Emerging future issues in HIV/AIDS social research. AIDS 2006 20:959–965

Gilbert L, Selokow T-A, Walker E (1997) Society, health and disease. An introductory reader for health professionals. Sigma Press, Pretoria, p. 97, 111

Gleick PH (2004) The world's water 2004–2005: The biennial report on freshwater resources. Island Press, Washington

Hooper E (2000) How did AIDS get started? S Afr J Sci 96(6):265–269

IRC (International Water and Sanitation Centre) (2006) Progress, lessons and the way forward. Available (online) at website: http://www.irc.nl/page/31686

Janse van Rensburg E (2000) The origin of HIV. S Afr J Sci 96(6):267–269

Jayne TS, Chapoto A, Byron MN, Hamazakaza P, Kadiyala S, Gillespie S (2006) Community-Level Impacts of AIDS-Related Mortality: Panel Survey Evidence from Zambia. Paper presented at the Principal Paper session, "HIV/AIDS and Rural Food Security in Africa", Allied Social Sciences Association Annual Meeting, Boston, 6–8 Jan 2006, p. 1

Kamminga E, Wegelin-Schuringa M (2005) HIV/AIDS and Water, Sanitation and Hygiene. IRC International Water and Sanitation Centre, Delft. Available (online) at website: http://www.irc.nl, p. 2, 11, 17

Kgalushi R, Smits S, Eales K (2003) People Living with HIV/AIDS in a Context of Rural Poverty: The Importance of Water and Sanitation Services and Hygiene Education: A Case Study from Bolobedu (Limpopo Province, South Africa). Mvula Trust, KIT and IRC International Water and Sanitation Centre, Delft, The Netherlands

Kuppusamy B (2005) Government blasted for bold steps. In: Asiafrica Features. Available (online) at website: http://www.aidsasiafrica.net

Lundberg M, Over M, Mujinja P (2002) Sources for financial assistance for households suffering an adult death.World Bank ResearchWorking Paper No 2508. December, Washington. Available (online) at website: http://wdsbeta.worldbank.org

Marais H (2005) Buckling. The impact of Aids in South Africa. University of Pretoria, Pretoria, p. 7–8, 25

Mujeeb S (1999) Human right violations of PLWA/HIV by their family members. Posting to SEA-AIDS. Available (online) at website: www.hivnet.ch:8000/asia/sea-aids

Myer L, Rodney IE, Ezra S, Susser ES (2004) Social epidemiology in South Africa. Epidemiol Rev 26(1):112–123

Nardi PM, Bolton R (1991) Gay-bashing, violence and aggression against gay men and lesbians. In: Baenninger R (ed) Targets of violence and aggression, Elsevier, North-Holland

Obi CL, Onabolu B, Momba MNB, Igumbor JO, Ramalivahna J, Bessong PO, van Rensburg EJ, Lukoto M, Green E, Mulaudzi TB (2006) The interesting cross-paths of HIV/AIDS and water in Africa with special reference to Southern Africa. Water SA, (July 2006) 32(3):323–343. Available (on line) at website: http://www.wrc.org.za/downloads/watersa/2006/Jul%2006/1955.pdf

Phaswana-Mafuya N (2006) Hygiene status of rural communities in the Eastern Cape of South Africa. Int J Environ Heal R (Aug 2006) 16(4):289–303

Phaswana-Mafuya N, Peltzer K (2005) Perceived HIV/AIDS impact among staff in tertiary institutions in the Eastern Cape, South Africa. J des Asp Sociaux du VIH/SIDA 2(2):277–279

Potter A (2006) Water, Sanitation and HIV/AIDS. Available (online) at website: http://www.mvula.co.za/page/539

Ramasar V. Erskine S (2002) HIV/AIDS: Africa's development crisis? In: Baijnath H, Singh Y (eds) Rebirth of science in Africa. A shared vision for life and environmental science, Umdaus Press, Hatfield, p. 22, 24

South African Medical Research Council (2005) Annual Report 2005. Available (online) at website: http://www.mrc.ac.za/annualreport/annual.html

Sufian S (2000) HIV/AIDS in the middle East and North Africa: A primer. In: Middle East Report. Available (online) at website: http://www.merip.org/mer/mer233/sufian.html

Tladi B, Baloiy T, Schreiber-Kaya A, Mathekgana M, Mangold S, De Klerk T, Winde F (2002) State of the environment report. NorthWest Province, South Africa. Available (online) at website: http://www.nwpg.gov.za/soer/FullReport/nw glance.html

UNAIDS (2003) Progress Report on the Global Response to the HIV/AIDS Epidemic. Geneva: UNAIDS. Available (online) at: www.unaids.org/ungass/en/global/ungass00 en.htm

UNAIDS (2006a) Aids Epidemic Update. Geneva: UNAIDS. Available (online) at: website: http://www.unaids.org, pp. 5–6, 19–20

UNAIDS (2006b) From crisis management to strategic response. In: 2006 Report on the Global AIDS Epidemic. Available (online) at website: http://www.unaids.org

UNDP (United Nations Development Programme) (2006) World water and sanitation crises urgently need a Global Action Plan. Available (online) at website: http://content.undp.org/go/newsroom/novembr-2006/hdr-water-20061109.en

UN-HABITAT (United Nations Human Settlements Programme) (2006) HIV/AIDS Checklist for Water and Sanitation Projects. Available (online) at website: http://www.unhabitat.org, p. 8

UNSGAB (United Nations Secretary General Advisory Board) (2006) Africa Dialogue. Issues Note on the Hashimoto Action Plan in the Context of the African Water (AMCOW) Agenda

van Dyk A (2001) HIV/AIDS care and counselling: A multidisciplinary approach. Pearson Education, Cape Town

van Wijk C (2003) HIV/AIDS and water supply, sanitation and hygiene. In: WELL – Resource Centre Network for Water, Sanitation and Environmental Health. Available (online) at website: http://www.1boro.ac.uk/well/resources/fact-sheets/fact-sheets-htm/hiv-aids.htm

Walker L, Reid G, Cornell M (2004) Waiting to happen – HIV/AIDS in South Africa. Lynne Rienner Publishers, London

Whiteside A, Sunter C (2000) Aids: The challenge for South Africa. Human & Rousseau, Cape Town, p. 36, 37, 44

Williams A (2005) Islam, gender, and reproductive health: Part 6 of 6. In: Environmental Change and Security Program. Available (online) at website: http://www.wilsoncenter.org/index.cfm?topic id=1413&fuseaction=topics.event summary&event id=146540

World Bank (2006) At a glance. In: News and Broadcast – Water. Available (online) at website: www.worldbank.org/water

Irrigation and Water Policies in Aragon

José Francisco Aranda-Martín

Introduction

The major advances of irrigation in Spain, and within the autonomous community of Aragon, basically took place during the twentieth century, the second half in particular. Nevertheless, major developments in Aragon that were included in plans that date from the beginning of the twentieth century – basically the Bardenas System (www.cgbardenas.com) and the Upper Aragon Irrigation Systems (www.cg-riegosaltoaragon.es) – have yet to be implemented.

As a result of social and economic changes, as well as development of new paradigms, opposition has grown to the development of new intensive irrigation systems. Currently the emphasis is on social irrigation and the modernisation of existing systems in order to make water use more efficient.

Not all of Spain's autonomous communities face the same situation. Some have already reached a level of implementation of irrigation schemes that meets most of their aspirations, or at least they will have done so by 2012. Aragon, however, is far from having achieved that goal. Over the last 14 years, major delays have built up over the development of infrastructure on which all of the community's political forces agreed in the so-called Aragon Water Pact of 1992 (Resolución aprobada, 1992).

The backlog in the development of infrastructure has been mainly a result of heated social conflicts. The need to achieve social accords that would provide political backing for some projects, while discarding others, became obvious and led in 2004 to the Aragon Principles of Water Policy that established a new strategy for water developments in the autonomous community (Bases de la Política del Agua en Aragon, 2007).

J. F. Aranda-Martín (✉)
Department of Planning and Development, Water Institute of Aragon, Capitán Portolés, 1-3-5, 9 floor, 50005 Zaragoza, Spain
e-mail: faranda@aragon.es

These principles included major agreements that pointed the way forward. In straightforward terms, they can be summed up as the following:

- Complete the new irrigation projects in the major areas of general interest.
- Modernise existing irrigation schemes in order to increase efficiency in the use of water.
- Where necessary, abandon irrigable areas that are excessively saline.
- Promote the use of modern technology, along with research and development, and technology transfer.
- Combat nonpoint source pollution.
- Promote social irrigation. ('Social irrigation' refers to small schemes that contribute to the social cohesion of a sparsely populated area. The term also applies to schemes that support crops in areas of low hydric stress, especially such Mediterranean crops as almonds and olives whose future is increasingly uncertain but that have an ecological value of great magnitude).

All of the above principles are to be observed under rigorous criteria of social, economic and environmental viability and in the knowledge that any decisions on new irrigation zones should not be taken as a function of conjunctural policies.

Irrigation Developments

The twentieth century saw huge advances in irrigation. When it began, the area under irrigation in Spain was – as it basically had been for a very long time – 1.2 million hectares and in Aragon, 172,000 hectares. By the turn of the century these figures had risen to 3.7 million hectares in the case of Spain and 474,613 hectares in Aragon. These advances were made despite the low implementation of various National Water Works Plans, from the so-called Gasset Plan of 1902 (Plan Nacional de Canales de Riego y Pantanos, 1902) and its successive modifications, to the National Plan of 1933 (the Lorenzo Pardo Plan) (Plan Nacional de Obras Hidráulicas, 1934) to the later socio-economic development plans.

Several factors lay behind this development. They included the social and political situation of the time, an economic model in which major emphasis was placed on the primary sector within the overall national economy and the need to produce food for a hungry population. These factors combined with an arid climate in which non-irrigated land was shown to be incapable of supplying sufficient food to meet the needs of the population.

The successes achieved throughout the twentieth century, the drastic changes in Spanish society in general and Aragon's in particular in both social and economic terms, along with new approaches towards development, and environmental and social issues, have led in recent years to a tremendous change of direction in the collective stance on irrigation and the development of new projects. All that despite the failure to conclude the Bardenas irrigation project that first appeared in the 1916 Works Plan (Plan Extraordinario de Obras Públicas, 1916), or the

Upper Aragon irrigation zone whose regulatory law saw the light in 1915 (Ley de 7 de enero, 1915).

Nowadays, there appears to be consensus within society on the need to make rational use of water, reduce nonpoint source pollution, and modernise the present irrigated zones to improve its efficiency and provide them with new technologies to improve the social aspects of irrigation, so making them more attractive to the new generations. Nor are there doubts of the development of new irrigation zones of a social nature, in backward areas, many of them with insufficient water. The aim of these projects is to ensure the continuation of traditional crops of each of the zones and at the same time to contribute to the consolidation of the social fabric that still prevails in huge areas.

Where consensus breaks down, however, is over new developments of highly technified irrigation zones that cover major areas. Controversies emerge on whether the areas in question are included in new plans or whether – as in the case of Bardenas and Upper Aragon – it is a question of completing long-established existing plans. In all these cases, the talk is of a moratorium, to consolidate work that is already under way, holding off on new irrigation areas until new water infrastructure is in place, or even providing subsidies for the abandonment of irrigable land. Some critics blame Aragon's new irrigation areas for all the ills that beset conservation of the aquatic environment in such singular regions as the Ebro Delta.

These changes in society's attitudes have been reflected in regulations within the 2002 National Irrigation Plan (Real Decreto 329/2002, 2002), which mark a change in the model from one based on major expansions to another that seeks to conclude the projects already underway on the basis of previous plans, while making the modernisation and consolidation of existing projects the priority.

More recently, in 2006, a further step was taken in this direction with the publication of Royal Decree 287/2006 of March 10 (Real Decreto 287/2006, 2006). The decree regulates urgent projects to improve and consolidate irrigation areas with the aim of achieving sufficient water savings to palliate the impact of drought. It provides for action to be taken over 866,898 hectares and estimates an annual water saving of 1,162 hm^3 in accordance with the concerns of more advanced societies on the need to improve the quality of both surface- and groundwater, as well as conserve environmental values on a basis of sustainable use of the resource as expressed by the European Union in the Directive 2000/60/EC of the European Parliament and of the Council, the Water Framework Directive (WFD) (Directiva 2000/60/CE del Parlamento Europeo y del Consejo, 2000).

The WFD, recognising that the community's waters are under growing pressure from continuously growing demand for good-quality water for all purposes, understands that quantitative control is a factor that guarantees good water quality. As a result, it urges the establishment of quantitative measures the objective of which is to prevent further deterioration of aquatic ecosystems while offering protection and improvements through the promotion of sustainable water use on the basis of the long-term conservation of available water resources, protection and improvement of the aquatic environment, reduction in the dumping or spilling of toxic waste and loss of vital substances, ensuring the progressive reduction of contamination of groundwater and preventing pollution of freshwater, and reducing the impact of floods and droughts. The aim of these measures is to ensure sufficient supplies of good-quality surface- and groundwater as required for sustainable, equitable use, while significantly reducing groundwater pollution.

One might therefore think that in Spain and other Mediterranean countries, where agriculture is by far the economic sector that uses most water for production, the time has come to put an end to the expansion of irrigation, one of the factors that contribute to nonpoint source pollution. Meanwhile, however, farmers are calling for limitless increases in irrigation that go far beyond what is needed to achieve sustainability.

With these viewpoints as the background, it is interesting to analyse, in general terms to begin with, the current world situation and compare it with the singular circumstances of Aragon – with special reference to the Ebro Basin, which accounts for 88.2% of the territory and 91.4% of the population – before going on to evaluate what the government and parliament of Aragon have outlined for the future of the irrigation policy within the Principles of Water Policy in Aragon.

The Current Situation

At the international level, the June 2008 World Summit in Rome of the United Nations Food and Agriculture Organization (FAO) voiced concern at the need to increase food production by 60% by 2030 in the face of a growth in the world's population from 1.6 billion in 1900 to 6.6 billion in 2007, and amid projections that it could possibly increase to 7.2 billion in 2015, surpass 8 billion in 2030 and reach 9 billion by 2050. Moreover, the increase would occur amid a scenario of growing demand for farm products for uses other than food, such as biofuels. Meanwhile, Spain's population grew from 18.8 million in 1900 to 44.7 million in 2005 and Aragon's over the same timeframe from 928,000 to 1.277 million.

According to the FAO, some 277 million hectares are irrigated worldwide (FAO Statistical Yearbook, 2007). This represents 17.99% of all arable land (1.54 billion hectares) and by 2030 the amount of land under irrigation in the developing countries alone will increase by more than 40 million hectares. Current data for Spain show 17,844.192 million hectares under cultivation and 3.7 million hectares (20.88%) of irrigation. The corresponding figures for Aragon are 1,703,993 hectares of which 474,613 hectares (27.8%) are irrigated (Anuario de Estadística Agroalimentaria, 2006).

Taking all these data into consideration, it appears that an expansion of the area under cultivation will provide only a very small proportion of the necessary increases in agricultural production. The remainder will have to be based on research, technological development, innovation and technology transfer that increase agricultural productivity, and expansion of irrigation, at all times on the basis of sustainable resource management.

All the above has to be achieved without forgetting the strategic nature of agriculture and its contribution to food security, or the difficulty of adapting it to the 'just in time' concept of the management of industrial reserves. In farming, reaction times are longer, and only planned production management can provide palatable results. The European Union's recent abolition of reserves has proved to be a genuine failure. Nor have the new tendencies in globalisation backed by the World Trade Organization, and their influence on the European Union's agricultural policy, added weight to proposals to abandon all projects to expand irrigation; rather, they have shown that decisions on new structural transformation projects in arid zones must never be based on conjunctural policies. And, as we shall see, Aragon's arid zones have annual rainfall of less than 350 mm, while potential evapo-transpiration is more than 1,300 mm per year and the production differential of dry material, in accordance with the Turc index, ranges from 11.4 in non-irrigated areas to 42 where there is irrigation.

Aragon Within Spain

Spain's political regime is extremely decentralised in favour of the regions (known as autonomous communities, of which Aragon is one). Each autonomous community has full authority over agricultural, rural development, environmental and other policies relating to water, while the central government has the full authority of providing the model for water management units in basins that encompass more than one autonomous community.

Some data will provide a better understanding of Aragon's situation. The autonomous community of Aragon, in north-eastern Spain, covers $47,700 \text{ km}^2$, divided into the provinces of Huesca ($15,600 \text{ km}^2$), Zaragoza ($17,300 \text{ km}^2$) and Teruel ($14,800 \text{ km}^2$). The population of 1,296 million is divided into Huesca (220,000), Zaragoza (932,000) and Teruel (144,000). More than half of the total population lives in Zaragoza city and its environs, meaning that the rest of the territory is very sparsely populated.

The legal framework, as it relates to water in the Aragon autonomous community, was established in 2001 by Law 6/2001 on Planning and Participation in Water Management in Aragon. The Principles of Water Policy in Aragon were developed based on the mentioned law (Bases de la Política del Agua en Aragón, 2007). This document, which provides a guide to planning, will form a necessary reference further on in this chapter as we refer to the perspectives for the future of irrigation in the community. It refers to Aragon's territory, its climate and water resources in the following terms:

- Its territorial diversity is probably Aragon's most significant characteristic, including such profoundly differentiated zones as the Ebro Valley, the Somontanos (Foothills), the Pyrenees and the Iberian Mountains (Fig. 1)
- The most immediate manifestation of this diversity is altitude, the decisive factor in land occupation; more than 60% of the province of Teruel is located above

Fig. 1 The Ebro Basin and Aragon's location within the Ebro, Jucar and Del Tajo basins
Source: Ebro Hydrographic Confederation, www.chebro.es

1,000 m, compared with less than 30% in Huesca and a little more than 6% in Zaragoza (Fig. 2)
- Climate is the factor that determines water resources and the form in which they are used; in this case, too, there are sharp territorial variations
- Similar differences are to be noted in precipitations which, in the Ebro Valley, are less than 350 mm (mean annual), only about half the national average, while in the Pyrenees they exceed 1,700 mm, much of it snow

Mean precipitation in millimetres for 1971–2000 in Aragon's three provincial capitals is shown in Table 1.

Table 2 provides data on the climate of several representative localities distributed throughout Aragon. Together with Figs. 3 and 4, it presents the hydric balance, potential evapo-transpiration (PET), precipitation and temperatures:

- As a general rule, the features that define Aragon's climate are the aridity that leads to a generalised hydric deficit; wide variations from year to year; continental extremes with major differences among the seasons in terms of temperature and precipitation and the intensity and frequency of winds. All these factors contribute to the difficulties of farming in non-irrigated zones.
- This climatic complexity combines with other hydrological factors to create profound diversity in the distribution of water resources on each side of the main river, the Ebro.
- Flows from the left bank of the Ebro contribute more than three-quarters of Aragon's total resources, while those from the right bank do not amount to even a fifth. In other words, in terms of resources per square kilometre, run-off from the left-bank flows is some five times greater than that from the others.

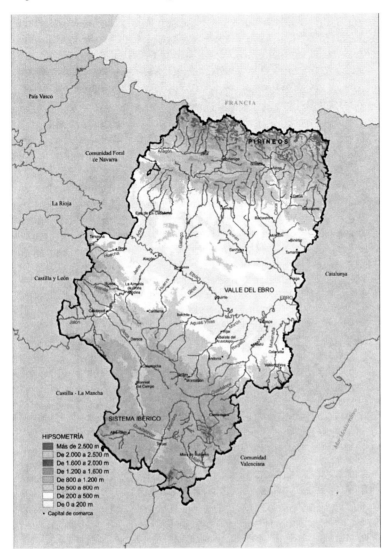

Fig. 2 The altimetrics of Aragon

Table 1 Mean monthly and annual precipitation for the period 1971–2000 in the three capital cities of Aragon

	Total	Jan	Feb	Mar	Apr	May	Jun	Jul	Aug	Sep	Oct	Nov	Dec
Zaragoza (mm)	318	22	20	20	35	44	31	18	17	27	30	30	23
Huesca (mm)	535	39	32	34	53	62	47	20	38	54	54	50	51
Teruel (mm)	373	17	14	19	36	56	43	30	40	36	42	22	20

Source: Agencia Estatal de Meteorología de España, http://www.aemet.es/es/nuevaweb

- Two other river basins within Aragon, the Júcar and Tajo, contribute in practical terms the remaining 5% of Aragon's resources, though their combined run-off is some three times less than that of the left bank of the Ebro.
- All these features, combined with the seasonality and inter-annual irregularity, explain why reservoirs have formed the traditional basis of Aragon's water use.

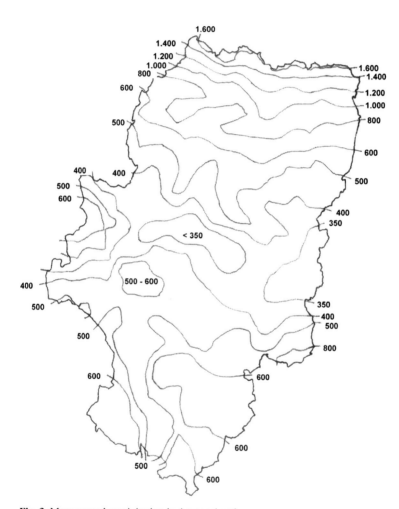

Fig. 3 Mean annual precipitation in Aragon (mm)

Table 2 Precipitation, temperature, evapo-transpiration and hydric balance in several cities representative of Aragon

Locality	Altitude (m)	Balance (mm)	PET (mm)	Precipitation (mm)	Max temp (°C)	Min temp (°C)	Mean temp (°C)
Jaca	818	−454.20	1,235.08	780.89	16.95	5.03	10.99
Sabiñanigo	780	−121.59	974.19	852.61	16.72	4.19	10.45
Ainsa	869	−116.45	946.76	830.31	17.71	6.50	12.10
Benabarre	782	−650.53	1,268.15	617.61	18.96	6.83	12.89
Sos	654	−495.40	1,114.89	619.49	18.49	7.42	12.96
Ejea	346	−712.48	1,165.72	453.24	19.52	8.59	14.06
Barbastro	341	−792.77	1,262.44	469.67	20.42	8.31	14.36
Monzon	279	−828.61	1,217.76	389.15	20.91	8.64	14.77
Sariñena	281	−840.44	1,245.44	405.01	20.60	8.72	14.66
Fraga	118	−886.05	1,259.08	373.03	21.00	8.81	14.90
Huesca	488	−681.11	1,212.60	531.49	19.81	7.89	13.85
Zaragoza	200	−855.81	1,189.92	334.10	20.41	9.09	14.75
Caspe	152	−934.66	1,240.27	305.61	21.86	10.15	16.00
Alcañiz	381	−820.64	1,196.72	376.08	19.78	9.19	14.49
Belchite	440	−853.16	1,164.47	311.31	20.55	9.62	15.09
Tarazona	480	−626.92	1,051.21	424.29	18.43	8.35	13.39
Almunia	366	−880.13	1,249.12	368.99	21.05	8.66	14.85
Calatayud	536	−965.61	1,311.64	346.03	20.44	6.49	13.47
Daroca	797	−761.52	1,175.48	413.96	18.96	6.07	12.51
Calamocha	884	−718.91	1,157.33	438.42	18.28	4.52	11.40
Cantavieja	1,200	−161.90	788.75	626.85	15.79	5.18	10.49
Teruel	915	−890.76	1,263.45	372.69	19.67	4.62	12.15
Albarracin	1,171	−538.31	1,074.93	536.61	15.88	3.28	9.58
Sarrion	991	−615.81	1,103.59	487.78	18.22	6.69	12.46

Source: Agencia Estatal de Meteorología de España, http://www.aemet.es/es/nuevaweb

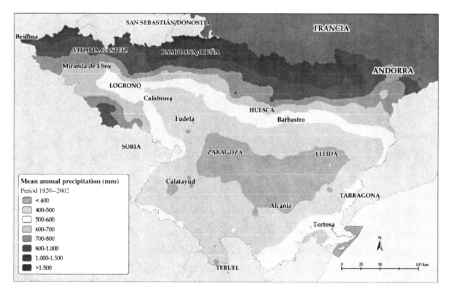

Fig. 4 Mean annual precipitation in the Ebro River Basin, 1920–2002
Source: Ebro Hydrographic Confederation, www.chebro.es

Aragon and the Water Pact

Throughout history, and given the responsibilities of the communities mentioned before, not all autonomous communities have been able to develop irrigation at the same pace. That will be even more evident in 2012 when the projects that were launched on the basis of the 2002 National Irrigation Plan come on stream. The majority of the autonomous communities that share the Ebro Basin have launched projects that will soon provide them with the irrigation they need to meet most of their aspirations. However, Aragon remains a case apart where, according to the National Irrigation Plan, projects are expected to increase irrigated area only by 17%, a figure that falls far short of the community's aspirations as expressed in various planning documents.

Far from achieving a healthy balance, Aragon continues to be stuck with a deficit in the development of infrastructure that as far back as 1992 led the region's political forces to state the community's aspirations on the use of the 7,400 hm^3 at its disposal in the Aragon Water Pact (Resolución aprobada, 1992). The Pact adopted as its starting point 3,600 hm^3 of consolidated consumption, of which urban use accounted for 150 hm^3; the rest would be used to irrigate some 390,000 hectares of farmland. The Pact also included the need for 2,100 hm^3 for projects that would have raised the area under irrigation to about 600,000 hectares, while meeting projections of 205.9 hm^3 of urban demand, and leaving a strategic reserve for future developments of 850 hm^3, for a total amount of 6,550 hm^3.

Fourteen years after the pact, and even taking into consideration that it was merely a political agreement that acquired the status of a regulation only when it was incorporated into the 1998 Ebro Basin Hydrological Plan (Real Decreto 1664/1998, 1998) and later into the National Water Plan of 2001 (Ley 10/2001, 2001), most of the infrastructures that were proposed have not been realised, in many cases because of the major social conflicts caused by some of the projects, as well as the new irrigation areas related to them. The development pace has fallen far short of meeting what was planned in 2001 by the National Water Plan as approved by Law 10/2001.

Aragon and the Water Commission

With this as the background, the Ministry of Environment of the government of Aragon, through the Aragon Water Commission, pointed in 2004 to the need for social agreements that would lead to the implementation of political decisions on some infrastructure developments while ruling out others. These new strategies for water-related developments were spelled out in the Principles of Water Policy in Aragon (Ley 10/2001, 2001).

It is worth pointing out that the Principles of Water Policy emerged as the result of intensive efforts by the Aragon Water Commission. The commission's 65 members represent 17 groups that in turn reflect the full range of Aragon's social sensibilities. The groups include environmental and ecological associations; groups of people affected by water projects; the university; water users of all types, including farmers, industries, city dwellers and recreational and sports associations; the autonomous community and local authorities; and water experts. The fact that the authorities are in a clear minority is what provides the real value to a text in which the Water Commission accurately reflected the sensibilities of the society of Aragon and which in its final form was approved by Aragon's Council of Government on 24 October 2006 and by the Aragon Parliament on December 1 of the same year.

The Principles of Water Policy include major agreements on the direction that is to be taken in the coming years. The most important infrastructure projects have to be completed and, once they are, a new stage of projects of a social nature has to be launched, without abandoning – but even strengthening – the modernisation of existing irrigation zones in order to achieve greater efficiency.

The principles also emphasise efforts to combat nonpoint source pollution, research, development and innovation, as well as technology transfer.

From Aragon's point of view, this decision would point to a future that can be simply summed up in a few phrases: 'Complete the expansion projects in major irrigation zones of general interest', 'modernise the existing zones to make them more efficient', 'where necessary, abandon irrigable areas that are excessively saline', and promote 'technification' and 'social irrigation'. The concept of social irrigation goes beyond the provisions of the National Irrigation Plan (Real Decreto 329/2002, 2002) to include very small projects that would help to maintain a minimum of social cohesion in areas where population densities are in some cases lower than five inhabitants per square kilometre. It also includes low-volume irrigation zones in areas of little hydric stress, especially where the survival of such traditional Mediterranean crops as olive and almonds – whose ecological value is immense – is increasingly threatened by uncertainty.

Principles of Water Policy in Aragon

The Principles of Water Policy in Aragon were drawn up by the Aragon Water Institute in accordance with Law 6/2001 on Planning and Participation in Water Management in Aragon. The law defines it as an instrument of water planning that must put into effect, insofar as is possible, Aragon's water policy as represented by the majority positions of its complex and plural society expressed during more than 2 years of work on a specific proposal from the Aragon Water Commission. Following a period of public information in 2006, it was approved by the Council of Government and later by the Parliament of the autonomous community. The document affirms the major role that irrigation has played, and continues to play in Aragon's economy, as well as in balancing regional differences and in socio-economic development. It backs the consolidation of current irrigation zones but also the development of new ones, including those for social purposes, within parameters of environmental, economic and social viability that are not linked to any of the European Union's sectoral policies, on the understanding that irrigation contributes to the formation of social capital that transcends political conjunctures.

Thus the Principles of Water Policy in Aragon set the guidelines for a series of issues that relate to agriculture and irrigation, as well as those that relate to other water-related activities. These are given below.

Guidelines Regarding Agriculture

- Farming is more important in Aragon than in many other autonomous communities. Aragon contributes with 4.3% of Spain's farming output in terms of net value added, while its contribution to industry is 3.5% and to services 2.9%.
- The two pillars of Aragon's farming economy are irrigation and intensive livestock farming. The latter accounts for 52% of farm production while crops provide 44%. Crops are grown over 37% of Aragon's territory – 1.8 million hectares of which 25% is irrigated, though the irrigated land accounts for 60% of production.
- The relation between irrigation and stock-rearing is of special strategic importance, given Europe's severe deficit of high-quality vegetable protein and the contribution of Aragon's irrigated pastures to its reduction. This contribution to the food security of the animals, and hence of the population, must be taken into

account in comparing the value added by this type of farming and that of more intensive farming of fruit and vegetables.
- Aragon's farm sector, like that of other autonomous communities and other countries, has to be seen as an integral part of an agricultural and food chain of ever-growing complexity in economic, social and environmental terms. Public administrations are obliged to guarantee the production of sufficient food, in terms of quality and quantity, to meet the growing demand within acceptable conditions of sanitary security, sustainable use of natural resources and the opening of international markets.
- The Integral Plan of Demographic and Population Policy (Plan Integral de Política Demográfica, 2000) passed by the Aragon Parliament in 2000 made clear that the desertification of land in Aragon can only be halted by integrated policies for rural development that cannot be restricted to farming, tourism or environmental protection. Rather, these must be seen as some of the essential activities that allow for the consolidation of a modern rural environment based on a more complex and diversified economic fabric of which, however, modern and efficient farming is a vital ingredient.
- As a result, irrigation demands the maximum attention of public authorities because of its fundamental role as an activity that develops the territory of Aragon, while binding it together. It is, however, an activity that will have to move as swiftly as possible towards sustainable use of water resources.

The European Union's Common Agricultural Policy (CAP) and Its Influence on Irrigation

- Some 54% of Aragon's non-irrigated agricultural production – 57% in area terms – currently receives aid from the CAP, but aid is given to only 19% of irrigated production, or 53% of irrigated land. Irrigated zones in Aragon are, therefore, much less dependent on the CAP than are those without irrigation. As a result, irrigation in Aragon is unlikely to be very affected by the new policy guidelines published in 2006 or those that are likely to emerge in the future.
- Irrigation in Aragon's relatively low level of dependence on the CAP does not mean that the policy should be rigidly imposed on it, nor should it be concluded that the new aid regulations will not favour the continuity of the current irrigation zones.
- Given the CAP's susceptibility to change, any decision to link it to water policy implies severe political risks.

The Relationship Between Irrigation and Rural Development

The importance of irrigation to Aragon's economy is steadily being reduced by the inevitable advance of the service sector, but its profoundly positive impact on the rural development of an arid and intensely desertified autonomous community such as Aragon has never been placed in doubt. This is because regions where irrigation has a significant presence exhibit the following characteristics:

- Population densities are higher than those of non-irrigated regions, with higher growth rates and lower losses through emigration.
- Lower indices of aging, with a higher proportion of young people and women, as well as more job opportunities.
- Higher indices of population replacement and greater assurances that established activities will be maintained.
- In Aragon, irrigation has permitted a stabilisation of the rural population that would otherwise have been unthinkable. Likewise, it has helped to diversify farming – production of livestock now exceeds that of crops – and the development of agro industry and all related services.
- In this sense, irrigation is an essential factor in the conservation and settlement of population in the environment. Even allowing for the development of other activities within the rural economy, the stability of a major part of the population cannot be explained without the presence of irrigation. Priority must, therefore, be given to policies that strengthen and modernise family farms, the creation of areas of social irrigation, and encouraging people to stay on the land through successive generations.

The Future of Irrigation in Aragon and the National Irrigation Plan

Irrigation and Development

- By a very long way, irrigation is the leading water use in Aragon. Sustainable use of the region's water resources therefore depends very heavily on the compatibility between irrigation and protection of the environment. In this sense, the characteristics of irrigation in Aragon are highly favourable: it is extensive, it focuses on the provision of quality animal feed and it creates spaces that favour biodiversity. The main difficulty lies in ensuring the necessary water and progressive development towards reasonable cost recovery in accordance with the principles of the European Water Framework Directive.
- The introduction of irrigation is no longer the only or indispensable means of achieving rural development, whether in Aragon or in any other Spanish autonomous community of continental or Mediterranean agriculture. Increasingly,

irrigation must be seen not as the policy of a sector, but as an element within an overall land policy that involves several sectors.
- Aragon has an incalculable hydraulic capital up in its current irrigation zones. The principal objective must be to protect and consolidate that capital by means of a sustainable farming system and the modernisation of irrigation rather than simply extending it. That does not mean, however, that irrigation should not be extended where it is environmentally, economically and socially viable – criteria that ought to be applied whether irrigation is being promoted by the public sector or by private interests.
- The priority for existing irrigation zones is:
    - Modernisation of water infrastructure and management wherever hydraulic efficiency is low.
    - Ensure the necessary water resources in zones where supply falls short of demand.
- Ensuring sufficiency of water resources means increasing their availability from reservoirs, groundwater or the elevations in the main rivers and canals. Given that they aim to consolidate existing irrigation schemes, these measures must be given priority over others that would increase the availability of water for extensions or new schemes.
- Priority must also be given to modernisation. The National Plan foresees the modernisation of almost 285,000 hectares.
- Modernisation of such a quantity of land, including improvements in territorial distribution brought about by the concentration of smallholdings, must observe criteria of strict social and environmental objectivity consistent with rational planning.

National Irrigation Plan's Major Projects

The National Irrigation Plan falls short of meeting all of Aragon's needs, but it constitutes an essential instrument in ensuring the future of the farm sector because it includes measures that are vital to the consolidation and modernisation of existing irrigation schemes as well as for the introduction of new ones that will contribute to sustainable development. As a result, the following are necessary:

- The measures that the plan scheduled for the period up to 2008 have to be executed. To that end:
    - The new projects have to be implemented by the state administration as quickly as possible, and they must be accompanied by measures that ensure the necessary water resources without upsetting the balance of the current system of exploitation
    - Modernisation projects must also include a financial framework similar to that of the new projects and incorporating measures at the level of individual farms and holdings

- The measures to be taken beyond 2008 that have yet to be scheduled must be updated and planned on the basis of:
    - Active participation by those involved in the rural environment, with concrete proposals from the interested parties
    - The establishment of objective criteria for prioritisation that include the following elements at least
    - Risk of desertification
    - Social, territorial and environmental sustainability
    - Socio-economic viability
- Fulfilment of the financial commitments undertaken by the central government and that of the autonomous community

Social Irrigation Within the National Irrigation Plan

- The tendency to set a prudent limit on major new projects and to encourage small projects distributed over a wide area in general, and the least favoured zones in particular, is backed by the National Plan's programme of social irrigation
- Social irrigation is of special importance to Aragon because it implements rural development, including several measures in mountainous regions that contribute to the extension of farming and the conservation of eco-systems, along with the recovery of land affected by dam construction
- Another positive aspect of social irrigation projects is that they are applied in areas that are severely disfavoured and remote from major resource systems
- As with all the new projects, social irrigation measures programmed for after 2008 have to be planned on the basis of criteria of participation, sustainability and viability
- As a result, the various administration departments of the autonomous community have to coordinate efforts among themselves and with other administrations that have responsibility for water matters in order to ensure that all these measures are environmentally compatible

New Irrigation Schemes

- The autonomous community's stance on the need for new irrigation zones is a strategic option, conceived as an instrument of land regulation in order to achieve a better demographic balance. It is also based on the sustainability of water resources and on the possibilities for production that can be developed on the basis of the soil and climate.
- Aragon will therefore push for new projects, beyond the 2008 deadline, that conform to its strategic option. To that end, the government of Aragon has to study the new projects that meet these conditions and are environmentally compatible

before consolidating them, broadening its financial commitment, and incorporating them in future stages of the National Irrigation Plan.
- The viability of the irrigation projects and the priority given to their implementation have to be subject to the sustainability of water resources, the profitability of farming and the contribution made to achieving greater territorial balance within Aragon. As a result, the central government has to be asked to broaden the financial commitment it made in the National Irrigation Plan.

Water Pollution and the Code of Good Farming Practice

- The Ministry of Agriculture and Food (of Aragon) is to make an urgent evaluation of the degree of compliance with the Code of Good Farming Practice (Decreto 77/1997, 1997) approved in 1997, especially with regard to water pollution and the quality targets set in hydrological planning
- The conclusions of the evaluation have to be transmitted to the Aragon Water Institute so that, once their effectiveness has been proved, the appropriate measures can be taken in coordination with the relevant ministries of the Aragon government and the national water administration

Water Used Exclusively for Productive Activities

- Water is a fundamental resource for irrigation, industry, energy or tourism and has to be dealt with using criteria based on land regulation, economic policies and viability.
- The Principles of Water Policy in Aragon devote a whole section to target resources, growth forecasts, fiscal payments, modernisation of infrastructure, major resource systems for irrigation and small requirements, construction of new dams, the consolidation of the existing major irrigation systems, new schemes from the national plan and others that are loss making, or demand systems on the right bank of the Ebro River. For each subject the following was indicated:

Target Resources

- Together with the consolidation of under-resourced zones, the priority measures for irrigation in Aragon involve improvements and modernisation.
- The Aragon administration has to propose to the National Water Administration the establishment of target resources in accordance with the programmes for improvement and modernisation.

Forecasts of Growth in Consumption

- New schemes must be implemented as a consequence of the technical and environmental viability of the new water sources that are required, the profitability of each irrigable zone and the sustainability of a model of rural development in accordance with community policies.

The Payment of Fiscal Obligations in Proportion to Consumption

- Payment of dues regulated by the Water Laws in proportion to the volume really used is essential for the efficient use of hydric resources. This system favours the allocation of resources to the most productive crops and provides an incentive for reaping the maximum benefit from water management.
- The Aragon government will work with the Environment Ministry and hydrographic confederations in the planning and implementation of the alternatives that are needed so that the volume of water used by passive subjects is taken into account in the calculation of the Water Law's payments regime.

The Modernisation of the Public-Sector Infrastructure

- Modernisation of the public-sector infrastructure that transports water from reservoirs to irrigation areas, as well as of the infrastructure that transports the water within these areas – usually the property of the irrigators themselves – and of the irrigation systems included in the National Irrigation Plan, is vital in order to improve the efficiency of water use for irrigation in Aragon.
- Efficient use of water is regarded as the basic tool that permits irrigation to be compatible with 'sound ecological conditions'.
- Modernisation is one of the bases of improvements in efficiency, an aim that farmers believe to be necessary to the survival of current irrigation systems. It has also achieved a consensus in major sectors of our society.
- The greatest possible degree of support has to be promoted from the state and the autonomous community. To that end, priority will be given to aid for investments in areas declared to be in need of urgent action and for water-saving projects that will free up flows so that they can either fulfil their environmental mission or meet future demand from the irrigation systems.
- A detailed study is urgently needed of the transport capacity problems that can arise, or already exist, in the major irrigation systems because the canals were designed to provide much less water than is required for their current crops, or for those that can be foreseen on the basis of current projections. The economic and environmental costs of the increase in capacity of the canals that have problems will be estimated and taken into account when solutions are proposed to possible problems of regulation.

- The construction of lateral storage ponds should be promoted in order to rationalise water use.
- The causes of increased river pollution through farming have to be analysed. Solutions should be proposed that are based on a return to the natural state of the rivers, the encouragement of ecological farming, the abandonment of land that produces high levels of saline effluents, restrictions on the use of fertilisers and pesticides in the worst-affected areas, and the reuse of effluents, both when they reach rivers within the irrigation zones and when they form part of the recharge of groundwater through irrigation.
- The crop structure, together with system efficiency and the degree of reuse, is the basic factor that helps to determine the provision of water in any given zone.
- Transformation measures must include a rigorous analysis of their environmental, economic and social viability; in some cases the analysis will rule out the measures, in others it will make them budget priorities. The analysis should not only take account of the benefits of the zone in question but of any negative or positive impacts on neighbouring zones.
- All the measures for modernisation of the public-sector infrastructure that were in the Water Pact, but have yet to be implemented, must be undertaken or concluded as quickly as possible. The government of Aragon has to ensure coordination of these measures with those of the National Irrigation Plan.
- The Aragon government must also promote all the other measures for modernisation of the public-sector infrastructure that, though not in the Water Pact, form part of basin hydrological plans or are of interest as a result of the National Irrigation Plan's modernisation programme.

Cost Recovery

- The principle of the recovery of the costs of water used for irrigation in Aragon must be applied, in accordance with the European Water Directive as enshrined in Article 111 of the Water Law, taking account of the social, environmental and economic consequences of the various types of irrigated agriculture in Aragon and its conditions of geography and climate.
- Thus, the positive effects that each irrigation system has on territorial development and depopulation, the way in which it favours – or not – the conservation of eco-systems or the economic indicators of its profitability, and as a consequence its payment capacity, are factors that have to be taken into account in order to fix the degree of cost recovery applicable in each case, or rather to configure the regime of exceptions that the law provides for.
- These factors, and any others whose application may be opportune in accordance with the current water legislation, have to be weighed in homogenous fashion throughout Aragon and in all the autonomous communities; there can be no discrimination.
- Given the overwhelming importance for general interest in the exceptions regime that is to be established, the Aragon government has to be especially active in

drawing it up and applying it. The regime itself, according to the water legislation, has to be articulated by means of basin hydrological plans.
- To that end, the users' communities, associations and federations will be consulted by the Aragon authorities.

The Major Systems of Resources for Irrigation and Small-Scale Schemes

- Irrigation is Aragon's major water user. Given the quantities of water involved, Aragon's socio-economic development has to be made compatible with the good ecological condition of its rivers.
- Emphasis must be put on the reuse of water in the major irrigation systems as a way of contributing to a reduction in the spread of pollution caused by farming.
- The best way to amplify and create social irrigation projects is in the form of 'green patches'. The acceptability of projects will be based on the socio-economic benefits that they bring to the affected area. Parameters to be taken into account include the availability of water, scarcity of a suitable-sized area for irrigation within the municipality or district and the reordering of property rights if necessary.
- Attention must also be given to the supply problems faced by small-scale irrigation schemes that have no links to the major resource systems. The Aragon government has to adopt corresponding planning initiatives in these cases.

The Construction of New Reservoirs

- In order to define the necessary infrastructures for regulation at any given time, the broadest possible catalogue is needed of regulatory projects capable of storing water in the headwaters of the rivers as well as in the irrigable zones or zones of use.
- The catalogue has to include the dams and reservoirs included in the Water Pact, with the modifications that derive from the agreements of the Aragon Water Commission and a new list of possible regulatory projects, drawn up on the principles of participation, sustainability and viability. The list of dams will include information on costs as well as on social, economic and environmental impacts that provide an idea of their importance and to be taken into account in decision-making.

The Consolidation of the Major Irrigation Systems

- Consolidation of the major irrigation systems has to be achieved by ensuring water supplies, increasing the availability of resources wherever they are found to be insufficient.

- To that end, the first aim should be to satisfy, as soon as possible, the requirements of the various irrigation systems by whatever regulatory measures are deemed to be necessary – without prejudice to additional measures required for the smaller systems in the same basins.

The New Schemes Proposed by the National Irrigation Plan

- Extensions of existing systems and new projects linked to the major resource systems were foreseen by the National Irrigation Plan as a result of agreements between Aragon and the central government. In order to ensure supplies for these projects, measures must be taken to regulate and consolidate the resources within the Upper Aragon Irrigation Systems, while new regulatory measures will have to be taken in the basins of the Turia and Mijares rivers.

Other Irrigation Systems that Suffer from Water Shortages or Demand Systems on the Right Bank of the Ebro

- Some current irrigation systems, as well as other uses, that are unable to ensure sufficient supplies of surface or groundwater are unable to achieve additional resources of their own. Almost all of these irrigation systems and uses are located on the right bank of the Ebro, and measures have to be taken to resolve their plight.
- The aim of these measures will be not only to consolidate existing irrigation systems but also to create new ones, where that is deemed feasible on the basis of an analysis of the technical, economic, social and environmental requirements of each case.
- The possibility of supplying additional, external resources to these units should be considered when their own resources are insufficient to consolidate the existing uses.

Conclusions

In an international context, the United Nations Food and Agriculture Organization (FAO) leaves no room for doubt. 'There is a world food shortage that can only be alleviated if agricultural yields can be increased in significant and sustainable fashion. And that in turn will depend on an increase in the use of irrigation and improvements in water management, the FAO says' (FAO, undated).

The Water Framework Directive's proposals on the prevention of any further deterioration, and the protection and improvement of the aquatic ecosystems by promoting sustainable water use, does not oppose the increase of irrigation. Rather, these proposals aim to ensure sufficient supplies of surface or groundwater of good quality, as required for sustainable, balanced and equitable water use, while achieving significant reductions in pollution of groundwater.

Aragon's planning documents show that it has understood that irrigation is a strategic factor in the cohesion and development of its territory. As a result, public administrations are under maximum pressure to realise the potential of the irrigation systems in the understanding that they have to evolve as quickly as possible towards sustainable use of water resources, boosting family farming and its modernisation, the creation of social irrigation systems and the involvement and participation of new generations.

Aragon understands that its aspirations for the development of irrigation have yet to be met. Its priority objective is the consolidation of the existing irrigation zones within a sustainable farming system, while creating new irrigation projects wherever they are environmentally, economically and socially viable. This will mean increasing the availability of water resources by means of reservoirs, groundwater extraction or withdrawals from principal rivers or canals.

Within the analysis of environmental, economic and social viability, Aragon insists that the positive impacts of each irrigation area must be taken into account in terms of territorial development, loss of population, whether or not it is of a type that favours conservation of eco-systems and the economic indicators of the project's profitability, and hence of its capacity to pay.

References

Agencia Estatal de Meteorología de España, http://www.aemet.es/es/nuevaweb
Anuario de Estadística Agroalimentaria (2006) Ministerio de Agricultura y Alimentación, Madrid
Anuario Estadístico de la FAO vol 1/1 www.fao.org/statistics/yearbook/vol_1_1/index.asp
Bases de la Política del Agua en Aragon (2007) Boletín Oficial de Aragon n° 24 de 26 de febrero
Decreto 77/1997 de 27 de mayo, del Gobierno de Aragon, por el que se aprueba el Código de Buenas Prácticas Agrarias de la Comunidad Autónoma de Aragon y se designan determinadas áreas zonas vulnerables a la contaminación de las aguas por los nitratos procedentes de fuentes agrarias, Boletín Oficial de Aragon n° 66 de 11 de junio de 1997
Directiva 2000/60/CE del Parlamento Europeo y del Consejo de 23 de octubre de 2000, por la que se establece un marco comunitario de actuación en el ámbito de la política de aguas Diario Oficial n° L 327 de 22/12/2000 p. 0001–0073
FAO (undated) El Agua y la Seguridad Alimentaria, www.fao.org/docrep/006/j0083s/j0083s06.htm
FAO Statistical Yearbook (2007) vol 1/1, www.fao.org/statistics/yearbook/vol_1_1/index.asp
Ley de 7 de enero 1915 de Riegos del Alto Aragon. Gaceta de Madrid n°. 77, de 18 de marzo de 1915
Ley 10/2001 de 5 de julio, del Plan Hidrológico Nacional, Boletín Oficial del Estado n° 161 de 6 de julio de 2001
Plan Extraordinario de Obras Públicas Revista de Obras Públicas n°. 2143, de 19 de octubre de 1916
Plan Integral de Política Demográfica aprobado por el Gobierno de Aragon en Octubre de 2000, Boletín Oficial de las Cortes de Aragon n° 80 de 11 de octubre de 2000 y las Resoluciones sobre Política Demográfica aprobadas por el Pleno de las Cortes de Aragon en sus sesiones de 5 y 6 de abril de 2001 Boletín Oficial de las Cortes de Aragon n° 121 de 11 de abril de 2001

Plan Nacional de Canales de Riego y Pantanos de 1902 (Plan Gasset), aprobado por Real Decreto de 25 de abril de 1902, y publicado en la Revista de Obras Públicas, n°. 1.388, de 1 de mayo de 1902

Plan Nacional de Obras Hidráulicas de 1933 (Plan Lorenzo Pardo), Gaceta n°.182 de 1 de julio de 1934

Real Decreto 1664/1998 de 24 de julio, por el que se aprueban los Planes Hidrológicos de cuenca Boletín Oficial del Estado n° 191 de 11 de agosto de 1998 y ORDEN de 13 de agosto de 1999 por la que se dispone la publicación de las determinaciones de contenido normativo del Plan Hidrológico de la cuenca del Ebro, Boletín Oficial del Estado n° 222 de 16 de septiembre de 1999

Real Decreto Legislativo 1/2001, de 20 de julio, por el que se aprueba el texto refundido de la Ley de Aguas Boletín Oficial del Estado n° 176, de 24 de julio de 2001

Real Decreto 329/2002 de 5 de abril, por el que se aprueba el Plan Nacional de Regadíos, Boletín Oficial del Estado n° 101 de 27 de abril de 2002

Real Decreto 287/2006 de 10 de marzo, por el que se regulan las obras urgentes de mejora y consolidación de regadíos, con objeto de obtener un adecuado ahorro de agua que palie los daños producidos por la sequía, Boletín Oficial del Estado n° 60 de 11 de marzo de 2006

Resolución aprobada por el Pleno de las Cortes en su sesión de 30 de junio de 1992, con motivo del debate de la Comunicación de la Diputación General de Aragon relativa a criterios sobre política hidráulica en la Comunidad Autónoma de Aragon. Boletín Oficial de las Cortes de Aragon n° 40 de 7 de julio de 1992

Singapore Water: Yesterday, Today and Tomorrow

Teng Chye Khoo

Introduction

Singapore is a small island nation with a total land area of about 700 km^2, or 65% the size of the city of Zaragoza. However, the population in Singapore is 4.6 million, about 6.5 times that of Zaragoza. The result is a densely populated city that exerts great pressure on competing land uses such as housing, commerce, industry, transport, recreation, schools and universities and, on top of these, water catchments. With no natural aquifers or groundwater, Singapore is considered one of the water-scarce countries. UN studies (United Nations Educational, Scientific and Cultural Organisation, 2006) have ranked Singapore 170th among 190 countries in terms of fresh water availability. This is not due to the lack of rainfall (which averages 2400 mm/year) but rather because of our limited land to catch the rainfall and also because of our increasing population.

The challenge for the national water agency is to manage Singapore's water resources in an integrated manner so as to optimise these limited resources for sustainability. In 2006, the Third World Centre for Water Management (Tortajada, 2006) concluded that Singapore has been very successful in managing its water and wastewater because of its concurrent emphasis on supply and demand management, used water (wastewater) and stormwater management, institutional effectiveness, and creation of an enabling environment, which includes a strong political will, effective legal and regulatory frameworks and an experienced and motivated workforce.

This chapter traces the development of Singapore's water management approach – yesterday, today and tomorrow.

T.C. Khoo (✉)
Public Utilities Board of Singapore, 111 Somerset
Road No. 15-01 Singapore 238164, Republic of Singapore
e-mail: tckhoo@pub.gov.sg

Yesterday (1960s–2000)

Upon attaining independence in 1965, Singapore's GDP (S$8.8 billion vs S$229 billion in 2007) and population (1.9 million vs 4.6 million in 2007) were very low, as was the corresponding water consumption (350,000 m^3/day vs 1,560,000 m^3/day in 2007). In spite of these low consumption levels, the inadequate infrastructure and water supply management sparked water rationing in 1961 and 1963.

The Public Utilities Board (PUB) was formed in 1963 as a statutory authority to take over the provision and supply of electricity, water and piped gas from the then City Council. PUB went on to develop a number of water supply schemes in the Malaysian state of Johor (from the 1960s), under the terms of two water agreements with Malaysia, and in Singapore (from the 1970s/1980s). With these developments, water rationing was consigned to the past and water supply became less dependent on the vagaries of weather. A reliable and clean water supply system was thus secured.

At the same time, stringent pollution control strategies and measures were adopted and enforced. An important milestone in this respect was the Singapore River clean-up.

Until the 1970s, the Singapore River and its surrounding areas were heavily polluted due to the presence of commercial activities along the river banks, industries and farms located along the upstream tributaries and extensive street hawking without proper sanitary and sewerage facilities.

The challenge of cleaning up the Singapore River was a daunting task that could not be achieved through the efforts of any single entity. Therefore, in 1977, the Singapore government launched a 10-year programme involving various government agencies.

Through close coordination between the agencies responsible for land-use planning, housing, environment, trade and industry and transportation, the large-scale clean-up operation was conducted. Pollutive activities were relocated to new premises to other parts of the island. River beds were dredged. Muddy banks were transformed into scenic sandy beaches. Riverside walkways were built and landscaped. A massive exercise was implemented to resettle squatters, and more than 26,000 families were moved into proper public housing complete with sanitary facilities. These tasks were no mean feat, but their success was ensured by strong political will, oversight from the very top, and a common long-term vision for a better quality of life.

Ten years later, in 1987, with the integration of land-use and water planning, the Singapore River was transformed into a clean, unpolluted, stench-free body of water. Even aquatic life returned. It was something to make every Singaporean feel proud.

Moreover, from a water resource viewpoint, the clean-up was significant as it enabled the complete separation of drainage and sewerage systems and would set the stage for developing future water resources (see section on Local Catchments).

Today-Water for All: Conserve, Value, Enjoy

The dawn of the twenty-first century saw a major transformation for PUB. In 2001, PUB fully relinquished its energy portfolio and concurrently assumed responsibility not just for the water supply but also for sewerage and drainage. In this way, PUB was reconstituted to become the national water agency overseeing the holistic management of Singapore's water supply, used water (wastewater) and storm water. This also facilitated the development of NEWater (recycled water), which enabled PUB to short circuit the natural water cycle, thereby 'closing the water loop'.

This holistic approach extends to PUB's belief that the comprehensive management of water resources should comprise two dimensions – the hardware and the software. This is encapsulated in the six words of PUB's corporate tagline: Water for All: Conserve, Value, Enjoy.

'Water for All' refers to the supply strategy to ensure a diversified and sustainable supply of water for Singapore. This is the hardware aspect, which covers the four sources of water supply known as the 'four national taps': local catchment water, imported water, NEWater (high-grade recycled water) and desalinated water.

The second part of PUB's tagline – 'Conserve, Value, Enjoy' – describes water demand management. It reflects the involvement of the people–public–private (3P) sectors in conserving water, keeping water catchments and waterways clean and building a relationship with water so that everyone can enjoy Singapore's water resources. This engagement of multiple stakeholders through all levels of society is the software dimension of PUB's water management.

Hardware: Leveraging on Technology

Local Catchments

The Reservoir in the City – to maximise the collection of rainwater, about half of Singapore's total land area is currently used as water catchment. With the completion of the Marina, Punggol and Serangoon reservoirs, the water catchment will grow from the current half to two-thirds of Singapore's land area before the end of this decade.

Singapore was among the first cities in the world to obtain drinking water from estuarine reservoirs and urbanised catchments in the 1970s and 1980s. This prepared PUB well for establishing the Marina Reservoir in the heart of the city. Singapore's 15th reservoir is the result of earlier efforts to clean up the Singapore River, which now serves as an unpolluted source of rainwater that can be harvested for drinking. The Marina Reservoir will have a catchment that is more urbanised and larger than that of any other reservoir in Singapore. At 10,000 hectares (100 km^2), its catchment is equivalent to one-sixth the size of Singapore.

Many cities around the world are built around reservoirs, but Marina Reservoir is a reservoir created in the middle of a city. Dubbed 'the reservoir in the city', it will be formed at the confluence of five rivers by building a dam, the Marina Barrage, across the mouth of the Marina Channel. Completed at the end of 2008, the barrage will also act as a tidal barrier for flood control purposes, while the reservoir will offer numerous opportunities for lifestyle, recreational and sports activities in the heart of the city.

Imported Water

Singapore imports water from its neighbour, the Malaysian State of Johor, under two water agreements signed in 1961 and 1962. The 1961 Water Agreement will expire in 2011, and we will not need to renew it. The 1962 Water Agreement will expire in 2061, and Singapore can be totally self-sufficient then, if there is no new water agreement with Malaysia.

NEWater

NEWater is the jewel of Singapore's water supply diversification strategy. Produced from the reclamation of treated used water, it is now a reliable source of water supply for the commercial and industrial sectors, with the capacity to meet some 15% of Singapore's water demand in 2008.

The idea of NEWater started long ago in the 1970s. However, the high cost and unreliable nature of membranes at the time meant that water reuse was not viable. When membrane technology improved in the 1990s, the idea of water reuse was revisited. A full-scale demonstration plant was commissioned in 2000 to undertake extensive studies on the quality of reclaimed water and the reliability of membrane technology.

Leading water testing laboratories were engaged to carry out comprehensive physical, chemical and microbiological tests over a 2-year period. Chemical parameters of emerging concerns were also included. In all, some 190 parameters and over 20,000 tests were carried out on the reclaimed water.

In addition, an international panel of experts in engineering, biomedical science, chemistry and water technology was formed to provide independent advice on the water reclamation study and to evaluate the suitability of NEWater as a source of water for potable use. The panel regularly audited and reviewed the test results and the demonstration plant operations. After the extensive battery of tests and analysis, NEWater was certified to be consistently of high quality, well within USEPA requirements and WHO standards for drinking water. This led to the full-scale production of NEWater in 2003 as an alternative source of water.

One challenge PUB anticipated was getting the public to accept NEWater. Experiences in the United States, particularly in California, had shown that a comprehensive but clear and simple public education programme was needed. Therefore, a comprehensive public education programme was planned, involving briefings to the public, including community leaders, business communities and students.

In 2002, before unveiling NEWater, PUB considered two models of communication. The first involved informing the public of the NEWater study, followed by the public education programme, and finally, an announcement of the policy decision to have NEWater as the third national tap. The second model involved the same three steps in a different order: informing the public of the NEWater study, announcing the policy decision and then finally the public education programme. Through careful deliberation, PUB recognised the need for the public to understand what NEWater was before the policy decision could be revealed. PUB therefore decided to educate the public before making the announcement. This would not only prepare the ground psychologically but also provide an avenue for public feedback before implementing NEWater.

The media was a key partner in the successful acceptance of NEWater by the public. Recognising the role of the media in educating the public, PUB organised visits for newspaper and television journalists to the United States and Europe to witness and understand both planned and unplanned indirect water reuse. It was important for the media to be aware that water reuse was not new. In the United States, for example, water reuse was already practised by the Orange County Water District in Southern California and the Upper Occoquan Sewerage Authority in North Virginia. In other countries with long riverine systems, upstream communities had long been discharging treated used water back into the rivers. This was the case for cities along the Rhine and Thames rivers in Europe, the Mississippi in the United States and the Yangtze in China. After the media was well informed, PUB was able to engage them as partners to educate the wider public.

In this way, NEWater was also greatly publicised through the media. Furthermore, PUB set up a NEWater Visitor Centre as another means of continuous public education. Today, NEWater is widely accepted in Singapore as a reliable and safe alternative supply of water.

The primary use of NEWater is for direct non-potable use as its ultra-clean characteristics and purity render it well suited and cost-effective for use in water fabrication, various other industrial processes and air-cooling for commercial buildings. Since NEWater was launched in 2003, feedback from these users has indicated lower water consumption and as much as 20% savings in operational and maintenance costs. Besides non-potable use, a small amount is also pumped into local reservoirs to mix with rainwater for indirect potable use.

By using each drop of water more than once, NEWater effectively multiplies Singapore's water resources and contributes significantly to our water sustainability. PUB opened its fourth NEWater plant in March 2007. This plant, developed through a Design-Build-Own-Operate (DBOO) approach with Keppel-Seghers, is capable of supplying 32 mgd (145,000 m^3/day). Together, the four NEWater plants in operation can supply some 56 mgd (255,000 m^3/day) of NEWater, meeting some 15% of Singapore's water demand. The large increase in NEWater capacity has yielded significant operational efficiencies and economies of scale. This, coupled with the significant decrease in membrane cost over the years, has resulted in PUB's ability to lower NEWater prices from \$1.30/m^3 in 2003 to \$1.15/m^3 in 2005 and further to \$1.00/m^3 in 2007.

The fifth and largest NEWater plant, the Changi NEWater Factory, will also be built through a DBOO approach. The tender was awarded to Sembcorp Utilities in 2008. When Changi NEWater Factory is completed at the end of 2010, it will be capable of supplying 50 mgd (227,000 m^3/day) of NEWater. Hence, by 2011, the NEWater plants will have the combined capacity to meet some 30% of Singapore's water needs.

NEWater is the result of an important paradigm shift: viewing used water as an important resource, to be recycled and re-used. This is in contrast to the previous mentality, which was to treat wastewater for discharge into the sea. To reflect this mindset change, we introduced new vocabulary, replacing the term 'wastewater' with the new term 'used water'. This signals the new approach of water management and communicates to the public the need to look at water as a renewable resource.

With continuing economic growth and the likely corresponding increase in demand for NEWater from industrial customers, PUB sees the importance of R&D to ensure a sustainable water supply for the future. Currently, when NEWater is recovered from used water, a certain percentage that is not recoverable has to be discharged into the sea. To meet long-term demand for NEWater, PUB is investing in R&D to improve the recovery rate of used water to produce NEWater. In this way, we can ensure that water use is truly maximised.

Desalination

Besides NEWater, we also opened Singapore's first seawater reverse osmosis desalination plant in 2005. Built through a DBOO partnership with the private sector, the plant is one of the biggest seawater reverse osmosis plants in the world, with a capacity to supply 30 mgd (136,000 m^3/day) of desalinated water.

Deep Tunnel Sewerage System (DTSS)

Besides the four national taps, an integral part of Singapore's strategy to manage the entire water loop is the DTSS. The DTSS is a 48 km long network with tunnels 20–50 m below ground to divert used water from the eastern, northern and central parts of Singapore to a centralised treatment plant at the south-eastern end of Singapore. The gravity-driven DTSS provides Singapore with a sustainable solution for the collection and channelling of used water for treatment and reclamation into NEWater. Land, being a scarce commodity in Singapore, carries a high price tag in urban areas. With its centralised treatment plant, the DTSS will free up valuable land occupied by existing water reclamation plants and pumping stations scattered around the island. Such land can then be put to other uses, like housing and industrial and commercial developments.

Software: The People–Public–Private (3P) Approach

Besides expanding the water supply, PUB also takes pains to manage the demand side through a 3P approach involving the community, businesses and civic groups.

In the area of water conservation, a multi-pronged approach keeps water consumption levels in check. This consists of appropriate water pricing, mandatory water conservation measures, public education and efficient management of the water distribution system.

Pricing

In 1997, the water tariff was restructured in a step towards achieving the government's objective of uniform flat tariffs for both domestic and non-domestic users. In addition, to encourage water conservation and to reflect the higher incremental cost of additional supplies, the government began to levy the water conservation tax (WCT) from the first drop consumed. Prior to 1997, businesses paid higher tariffs for water and domestic users were exempted from WCT for the first $20\,m^3$/month. The implicit assumption behind this was that each household was entitled to a certain amount of water. By restructuring the water price, the PUB strove to drive home the message that water is a strategic resource, precious from the first drop. In line with this objective, domestic and non-domestic users alike have been paying the same tariff for the first block of $40\,m^3$/month since 2000. The tariff and WCT were also increased for households that consume more than $40\,m^3$/month (see Table 1)

With the closing of the water loop, potable water and used water should now be seen as a single product. Therefore, the government is currently considering moving towards charging the same price for both potable water and used water in the longer term.

Table 1 Singapore water tariffs

Tariff category	Consumption block	Before 1 July 1997				From 1 July 2000			
		Tariff	WCT	Total	WBF	Tariff	WCT	Total	WBF
	(m^3/month)	(¢/m^3)	(%)	(¢/m^3)	(¢/m^3)	(¢/m^3)	(%)	(¢/m^3)	(¢/m^3)
Domestic	1–20	56	–	56	10	117	30	152.1	30
	20–40	80	15	92	10	117	30	152.1	30
	Above 40	117	15	134.6	10	140	45	203	30
Non-domestic	All units	117	20	140.4	22	117	30	152.1	60

Mandatory and Voluntary Conservation Measures

PUB adopts a basket of mandatory and voluntary measures to promote water conservation. Mandatory measures include the use of low-capacity flushing cisterns and constant flow regulators. PUB has also embarked on community-driven public education programmes such as the Water Efficient Homes programme and the Water Efficient Buildings programme to encourage home owners and building owners to adopt good water-saving habits and measures. Results have been encouraging – per capita domestic consumption has decreased from 165 L/day in 2003 to 157 L/day in 2007 (see Fig. 1).

To help us achieve an even more ambitious target of 155 L/day, PUB has launched the 10-L challenge. This programme aims to encourage all Singaporeans to reduce their daily water consumption by 10-L through simple but effective water-saving tips, such as taking shorter showers, installing water bags in cisterns to reduce the volume of water used for flushing, and washing dishes in a filled sink instead of under running water. We believe that through education the public will grasp the importance of water and use simple, voluntary measures to conserve it.

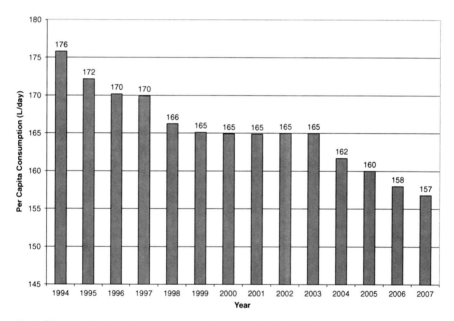

Fig. 1 Singapore's per capita water consumption from 1994 to 2007

Unaccounted for Water

In terms of unaccounted for water, we have spared no efforts to ensure that leaks in our water pipe network are kept to a minimum and water sold to customers is

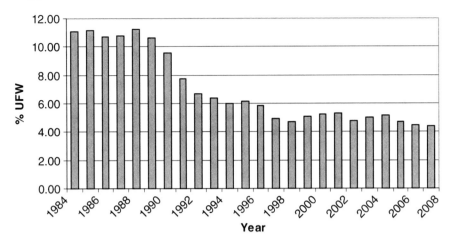

Fig. 2 PUB's Unaccounted for water from 1984 to 2007

accurately metered. This has enabled us to lower our unaccounted for water from about 10% in the mid-1980s to about 5% today (see Fig. 2). By reducing water losses, we help to keep water demand in check and there is less pressure to expand our water sources.

Engaging the Community

Beyond water conservation, PUB is encouraging people to value and enjoy water assets by opening up reservoirs for recreational activities. A public adoption programme called 'Our Waters' programme was also launched in 2005. Under this programme, organisations and community groups can adopt stretches of water and pledge to take care of the water resources by conducting clean-ups and river patrols and promoting greater awareness of water in the community. Public involvement will help ensure that these water assets are seen as a national resource to be cared for and cherished.

ABC Waters

To borrow Professor Malin Falkenmark's terminology, 'blue water' is the dominant source of water and a prominent feature in Singapore's highly urbanised landscape. We are surrounded by waterbodies and waterways – 32 rivers and what will soon be 17 reservoirs and 7,000 km of drains and canals. There is a need to keep this crucial source of blue water blue by involving civil society in its protection. Traditionally, engineering solutions entailed covering up drains and canals. This rendered such waterways invisible to the public, resulting in a sense of detachment between people and the waterways. Therefore, in 2006, PUB launched the Active, Beautiful and

Clean Waters (ABC) Programme, to transform utilitarian drains into beautiful and vibrant community spaces for recreational activities. This will not only enhance the overall quality of life for Singaporeans but also foster a greater sense of environmental ownership and awareness.

The efforts of yesteryear and today, addressing both the hardware and software, have enabled PUB to sail through rough seas. Over the past 40 years, Singapore's population has increased by two-and-a-half times, but the increase in water demand has been almost five times that. However, even in years of severe weather conditions such as El Niño in 1997–1998, Singaporeans still continued to enjoy an uninterrupted water supply.

Tomorrow–Water Management Beyond 2020

While PUB's efforts have been recognised internationally, there are still challenges to be addressed.

Challenge 1: Increasing Long-Term Water Supply

The first and most obvious challenge is meeting the increasing demand for water due to further GDP and population growth. Since independence, water consumption has grown significantly. Total water consumption in 2007 is almost five times that in 1965. Meeting the continuously increasing demand is one of the challenges PUB has to face in order to ensure both economic and social development.

Our water demand projection is based on the two main drivers of growth in water consumption, namely economic development (GDP growth) and population growth. As Fig. 3 shows, Singapore's population increased 2.5 times from 1965 to 2007, and GDP increased almost 26 times in the same period. As a result, water consumption increased dramatically between 1965 and 2007.

Water demand forecasting is possible only through active consultation with other national agencies in Singapore, such as the Economic Development Board, Ministry of Trade and Industry and the Urban Redevelopment Authority. Through integrated planning and tapping on the expertise of agencies involved in land-use planning and economic development, PUB is able to get the most accurate possible inputs for the water demand forecast model.

The water demand forecast serves two main purposes: capacity planning and cost and revenue projections. One important consideration is that, since non-domestic water consumption can be met with sources such as NEWater, the projections serve to outline NEWater capacity expansion to ensure that NEWater is used instead of potable water whenever possible. Ultimately, Singapore aims to increase the percentage of water demand met by NEWater to beyond 30%. However, increasing the use of NEWater for drinking purposes may require continued public education. This point was driven home by the failed referendum in Toowoomba, Australia, in July 2006, where residents voted against recycled water for potable use.

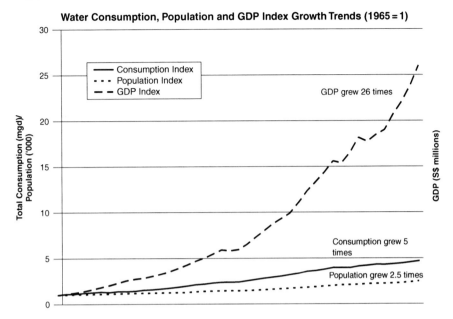

Fig. 3 GDP, Population and total water consumption growth (1965–2007)

Besides increasing the use of NEWater, PUB intends to address the supply side by harnessing more local water catchments. Unlike China or America, however, Singapore has limited rivers that can be dammed up to form new reservoirs. Because of such constraints, Singapore has to continue looking to technology for solutions. For example, PUB is exploring more advanced technologies such as variable salinity plants that can produce potable water from water bodies which are under the influence of the tides. We are also working with companies on projects like membrane distillation and ultrasonic sludge disintegration and achieving more recycling using membrane bioreactors.

Singapore also continues to invest heavily in R&D. A significant development in this area is the setting up of the National Research Foundation (NRF) in 2006 to drive R&D efforts in Singapore. The government has approved initial funding of S$5 billion for the NRF to undertake its work between 2006 and 2010. Some $330 million has been earmarked to promote R&D in the field of environmental and water technology sector. To map out the strategies and oversee growth in this sector, an Environment and Water Industry Development Council was also formed.

With the globalisation of the water industry, geographical barriers to technology transfer have now been removed. Singapore's local water expertise translates into a global solution; technology can be exported. It is Singapore's goal to be a Global Hydrohub, and there have been several significant developments in this area:

- General Electric Water signed an agreement with the National University of Singapore in September in 2006 to establish a world-scale R&D centre for water technologies. GE Water will invest S$130 million over 10 years, and the centre

will employ 100 top-tier researchers in water-related technologies to transform innovative concepts into real-world solutions that will be useful not only to Singapore but also to the rest of the world.
- Siemens Water Technologies plans to set up a S$50 million global water R&D and engineering centre in Singapore over the next 5 years to expand its Singapore location into a competence centre for water and used water technologies in the Asia-Pacific Region. Siemens Water Technologies also signed a memoradum of understanding (MOU) with PUB. Under the MOU, both parties will jointly work on exploring innovative solutions to meet technological needs through R&D, test bedding and early adoption of new technologies. For a start, Siemens Water Technologies will work with PUB on three R&D projects, mainly in the areas of water and used water treatment.
- PUB, the National University of Singapore and Dutch water consultant Delft Hydraulics signed an MOU to establish a regional centre of excellence for water knowledge with close industry linkages. The centre aims to embark on cutting edge research to serve the needs of Singapore, the region and beyond. The MOU marks the start of a Delft-Singapore alliance, paving the way for collaboration in multi-institutional and interdisciplinary research, information exchange and technology transfer relating to water management, hydraulic engineering and urban water cycles.
- Nitto-Denko has invested S$6 million to set up a water research and development centre in Singapore in 2008. The centre will be the first R&D centre set up by a Japanese water company in Singapore and will be located at WaterHub, PUB's research and development centre.

Challenge 2: Optimisation

The second challenge is to optimise our used water management system. To this end, PUB is studying how best to expand the DTSS system. The first possibility is to retain the existing water reclamation plants (wastewater treatment plants) in the western part of Singapore. Alternatively, a second DTSS could be constructed to serve the same region. As with the existing DTSS, this would free up valuable land for other uses. In addition, having centralised treatment facilities would mean greater economies of scale. Ideally, the extension of the DTSS would take place together with the implementation of newer technologies, resulting in a more efficient and cost-effective used water management system.

Challenge 3: Environmental Sustainability

Growing consciousness of how human activities impact the environment has led to a shift in the way PUB plans its projects. Therefore, the third challenge is including environmental sustainability as one of PUB's key objectives. With sustainable development as a goal, a more holistic approach is being taken in planning and implementing major projects. This includes carrying out environmental studies and putting in place environmental monitoring and management programmes for projects that impact the environment whether in terms of water quality or biodiversity.

PUB has also taken note of growing concerns on global warming. Together with the National Environment Agency of Singapore, we are studying the effects of climate change, for example, rising sea levels, changing rainfall patterns, temperature increases, and attempting to foresee their impact on water supply.

Technology has weaned Singapore from dependence on external water sources, but makes us more energy dependent. The escalating prices and uncertainty about the future availability of traditional energy sources are of particular concern. The production of NEWater currently requires five times as much energy as conventional treatment, while desalination requires a startling 20 times. Therefore, there is an urgent need to direct R&D towards low-energy, high-technology solutions.

In the area of seawater desalination, the Environment and Water Industry Development Council issued a request-for-proposal in 2007. Applicants were asked to propose innovative desalination solutions that consume 1.5 kWh energy or less per cubic metre of potable water produced from seawater. This is 50% more energy efficient, as compared to the advanced seawater desalination technologies currently available. The project was awarded to Siemens Water Technology in 2008.

Finally, by participating in global water events and associations, PUB strives to connect with the rest of the world in order to foster exchanges of ideas and insights into environmental sustainability with different countries.

The Road Ahead

Even with the most comprehensive plans and visions of a sustainable future, nothing can be accomplished without strong political will, good governance, effective implementation and a motivated workforce. Indeed, PUB has succeeded by working hard to be 'an institution which provides a near classic example of how public sector service can be made to perform efficiently and provide high quality water supplies at minimal waste' (World Bank Report, 1967–1989). But Singapore is not an isolated example. For instance, Professor Asit Biswas has noted the marked improvement in Phnom Penh's water supply system, where non-revenue water has dropped from more than 70% to about 8% today. The Asian Development Bank cites governmental support, civil society involvement, autonomy of the water operator and cost recovery as components of Phnom Penh's success. These factors are similar to those that form the basis of Singapore's water supply strategy.

Singapore has taken about 40 years to turn our vulnerabilities into strengths and attain self-sufficiency in water. Who knows what the next 40 years may hold? What we do know is what we would like to be – a city of the future, a city of gardens and waters. To achieve this, PUB needs to be constantly looking ahead and ready to embrace innovation, not just in technology, but in our whole approach to integrated water management. Only then can we be sure of sustainability beyond 2020.

References

Tortajada C (2006) Water management in Singapore. Int J Water Resour D (June 2006) 22(2):227–240

United Nations Educational, Scientific and Cultural Organisation (2006) Water – A Shared Responsibility (United Nations World Water Development Report 2.), Paris.

World Bank Report (1996) 'Water Supply and Sanitation Projects: The Bank's Experience – 1967–1989', Washington, D.C.

The Pace of Change in Seawater Desalination by Reverse Osmosis

Ian Lomax

Introduction

For many years desalination was regarded as a process the application of which was, on the whole, restricted to the Middle East and a few island communities around the world where rainfall or collection areas to harvest rainfall were insufficient to sustain the local population's needs.

The first generation of desalination equipment was generally a thermal-based, evaporative process, and water produced by this type of desalination process was regarded as being extremely expensive in comparison to the cost of potable water produced in areas having access to natural supplies of freshwater.

By the mid-1970s, a new process technology, reverse osmosis (RO), was becoming available to produce potable quality water from seawater or other brackish sources.

The Rapid Rate in Development of Seawater Reverse Osmosis

At this time, the process was still in its infancy, and the first seawater reverse osmosis (SWRO) plants were of low capacity, typically producing less than 100 m^3 of freshwater per day. The cost of each cubic metre of water produced from such plants was approximately US$2.40.

Table 1 shows the change in the typical cost of water production in the next 25 years.

By 1990, individual plant capacity had increased by more than two orders of magnitude with plants able to produce more than 40,000 m^3 each day. As plant capacity increased, the unit cost of water was reduced so that the typical cost of water produced by SWRO was about US$1.40 m^3. The reduction in the unit cost was

---

I. Lomax (✉)
FILMTEC (tm) Membranes, Dow Water Solutions, Dow
Deutschland Anlagengesellschaft mbH, P.O. Box 20, D-77834 Rheinmunster, Germany
e-mail: ilomax@dow.com

Table 1 Cost of desalinated water by reverse osmosis 1980–2007

Year	1980	1990	2000	2005	2007
$/m^3	2.43	1.40	1.00	<0.5	0.65

Source: Affordable Desalination Collaboration,
http://www.affordabledesal.com/home/news.html

achieved partly by economies of scale, but mainly by improvements in membrane technology and the introduction of energy recovery devices (ERD) that are able to utilise some of the pressure energy contained in the brine or waste stream from the SWRO unit, which had previously been discharged untapped.

Ten years later, plants of up to 100,000 m^3/day capacity were being built while the typical unit of water had decreased to around US$1.20/m^3.

Around 2001, improvements in ERD technology became available that enabled further reductions in the unit cost of water to be made to below US$1.00/m^3. By 2005 a plant in Ashkelon, Israel, was starting production which had a nominal production capacity of 340,000 m^3/day and an initial, published cost of water below US$0.50/m^3.

Since Ashkelon, many more plants have been proposed with capacities in the range of 100,000–500,000 m^3/day. Recently, the typical unit cost of water quoted for new, large plants has increased to a range of US$0.65-US$0.85/m^3.

To understand why the unit cost trend has reversed in this period, it is necessary to know something about how the cost is made up. Taken on a percentage basis, the cost of a cubic metre of water is approximately made up as shown in Table 2.

The recent increase in cost is largely due to increases in the cost of energy and the costs of some key materials of construction at the different sites around the world.

Despite these changes, there is a belief in some parts of the desalination community that the unit cost of water can be reduced by further reductions in the capital expense (CAPEX) and the energy required in the operating expense (OPEX) by the SWRO process. These further reductions in CAPEX and OPEX are expected to come from improvements in membrane technology, improvements in ERD unit efficiency and further optimisation of the SWRO process.

Table 2 Costs associated with water production

Capital	Power	Chemicals	Membranes	Labour	Total
~45%	~35%	~10%	~5%	~5%	100%

Source: Mickols, W.E., Busch, M., Maeda, Y., Tonner, J. "A novel design approach for seawater plants", International Desalination Association (IDA) 2005 World Congress on Desalination and Water Reuse, SP05-052, Singapore.

Future Developments

Reductions in the pressure required by SWRO are expected to come from the development of higher productivity membranes that will enable more flow to be produced for the same expenditure of energy. Also, the use of higher temperature feed water sources, such as the cooling streams from coastal power plants, could help to reduce the pressure needed to operate an SWRO process. Improvements to membranes and pre-treatment processes will also allow SWRO systems to be operated at higher recovery rates than are typically used today. Increasing the recovery rate of any SWRO system will allow more freshwater to be produced from the same capital investment in intake structure, pre-treatment plant and infrastructure.

If we go back to the historical time line in 1980, a typical SWRO system would convert 25–35% of the feed seawater into potable water. A modern plant such as Ashkelon is able to convert 45% of the feed into product, and it is becoming possible for plants to convert 55–60% in the future. A plant able to operate at 50% recovery rate will produce twice as much freshwater from the same quantity of feed water as a plant operating at only 25%. So it is becoming possible to potentially double the output of an existing, older plant for relatively little additional CAPEX.

Improvements in Membrane Productivity

One of the enabling factors of this potential to increase system recovery has been the remarkable improvement in the productivity of SWRO elements, as depicted in Table 3.

In 1985, a typical SWRO element would produce approximately 4,000 gallons (US) per day (GPD) of freshwater and would reject 99.4% of the salt. Currently, a typical element would produce roughly 8,000 gallons and reject 99.8% of the salt under the same conditions. This means that in a period of only 20 years, productivity has increased by a factor of two at the same time as the amount of salt able to pass through the membrane has been reduced by a factor of three, decreasing from 0.6 to 0.2%. This decrease in salt passage means that, in many cases, modern SWRO plants can meet current World Health Organization (WHO) and European Union (EU) guidelines for drinking water in a single-stage process.

Table 3 SWRO productivity improvement rates

Year	1982	1985	1990	1995	2000	2007	2007
Unit gpg	<4,000	4,000	5,000	6,000	7,500	8,000	12,000
% SP[a]	>0.6	0.6	0.4	0.3	0.25	<0.25	0.3

[a] At above-mentioned flow
Source: Filmtec Corporate product specification and data sheets

Two-stage or multiple-stage processes may still be required to meet specific limits on some ions such as chloride, if necessary for compatibility with old distribution system piping, or boron, if low levels are required to allow the water produced to be used for irrigation of crops such as citrus fruits. The need for a multi-stage process generally increases both the CAPEX and OPEX of the process.

These dramatic improvements in SWRO element performance and the development of ERD technology have brought about significant reductions in the amount of energy required to desalinate a unit quantity of seawater. Over a period of 20 years the power typically required to produce 1 m^3 of desalinated water has decreased significantly. In 1985, 6–8 kilowatt hours (kWh) (capital) would have been required. By 2000, a typical value would be around 4.5 kWh, and today the value would be in the range of 2–3 kWh. In extreme situations, this value has been reduced below 2 kWh, but in most commercial situations 2–3.5 kWh is more normal.

At this point it is probably necessary to stress that all of the costs for water production and the energy consumption values given above are only for the production of the water at the plant and do not include the capital and operating costs of pipelines or other distribution systems to get the desalinated water from the plant to the consumer.

Seawater Reverse Osmosis and the Environment

As the size of individual SWRO plants and the concentration of all desalination plants of all types in some regions have increased, so has the potential impact that such plants have on the environment. Many issues can be, and have been, raised about the location and positioning of major civil engineering infrastructure projects in coastal locations.

Apart from the issue of 'industrialisation' of coastal areas, the objections raised most frequently about the actual SWRO process seem to relate to the energy consumption and the potential effects of the intake and brine disposal on the ocean, both in the immediate vicinity and in the extended area where multiple desalination facilities may be concentrated in a relative small area of the coast.

The energy issue is, of course, a very valid one, but in objecting to some recent desalination projects that have been proposed around the world in such diverse locations as London, California and Western Australia, it seems as if the objections are being based on power consumption figures, such as 6 kWh/m^3, which are somewhat out of date, rather than the actual values of 2–3 kWh/m^3 that will be attained in a modern SWRO system. Unfortunately, the public, and in some cases 'expert', perceptions of the power consumption of SWRO lags behind the real situation, as is evident in Table 4.

Certainly, the removal of salt from seawater requires an amount of energy, but is this energy demand actually that high in relation to other activities? On average, I drive my car about 2 h/day. At a conservative estimate, my car requires approximately an average of 50 kWh of power during my journey, using about 100 kWh of

Table 4 Power consumption of SWRO plants

Source of power data	Old plants	Public perception	Expert perception	Current plants	Future targets
kWh/m^3	6–8	4–4.5	3–3.5	2–3.5	1.5–2.0

Source: Affordable Desalination Collaboration Results, Seacord, Coker, McHarg, Desalination and water re-use 07, 2006. Available at http://www.affordabledesal.com

power to get to work and back. In desalination terms that 100 kWh of power would produce something like 30 m^3 of freshwater.

The issues involving the effects of intake structures and the disposal of the brine created by a major SWRO plant are somewhat more complex. In relation to the intake, many concerns have been expressed, but they generally relate to the possible entrainment and destruction of fish or other marine life in the plant. Another concern is one relating to the possible modification of, or creation of, currents in the vicinity of the intake that may affect the immediate environment and the extended environment. For many proposed large plants, advanced hydraulic modelling seems to be able to provide good data on the potential impact of currents, and careful design of the intake structure, combined in some cases with suitable screens, appear able to minimise the risk to marine life in the vicinity of the inlet structure.

The concerns expressed about the potential impact of the brine relate to salinity, temperature, chemical additions, pH, oxygen depletion and other issues. The relative importance of specific issues may change with the location and the specific conditions of the environment into which the brine is to be discharged. A brine discharge from a typical SWRO plant can be expected to have a salinity that is between 1.5 and 2 times greater than the bulk seawater around it. Although it may be at twice the concentration of the brine stream, it is not actually adding any additional salt to the environment. The salt is contained in half the volume so that the total mass of salt returned to the sea in the brine stream is almost exactly the same as that removed from the sea in the feed stream. In this sense, the impact of the intake and disposal of water by RO is similar to natural evaporation, but only in a very limited area.

Unlike a thermal desalination process, the brine stream from a SWRO plant is typically about 1–3°C higher than that of the feed water stream entering the plant. The increase in temperature across an SWRO plant is principally due to heat gained as the feed flow passes through the high-pressure pumping stage. In a pre-treatment process, the seawater feed stream may be injected with a low dose of chlorine to help keep the intake structure clean. A coagulant, such as ferric chloride, may be added and a polyelectrolyte may also be added to assist with the flocculation and removal of particles on the filters downstream. Generally, the dosing of such chemicals is closely controlled, partly on cost grounds but also because excess dosing can result in the carry forward of iron and/or polyelectrolyte into the actual SWRO system where they can foul the membrane.

Most of the iron or polyelectrolyte that has been added is removed during the pre-treatment process. In some plants, the pH value of the feed seawater may be increased or reduced to help to either reduce the risk of the deposition of calcium carbonate scale or improve the rejection of some ions whose rejection by the membrane is affected by pH. Periodically, an SWRO plant may need to be cleaned either to remove fouling due to biofilm formation in spaces between the membrane surfaces or to dissolve sparingly soluble salts that may be precipitated from the brine solution as it is concentrated during the RO process. The bulk of the cleaning chemicals, usually based on a caustic detergent or relatively weak acid solution, are collected and treated before discharge to the drain. However, some traces may still be washed through the RO plant when it is put back into service.

Much environmental interest has centred on the recently commissioned SWRO plant located in Kwinana, Australia. The plant is situated in Western Australia on Cockburn Sound. Cockburn Sound, due to its location and special geographical features, is sensitive to the possible effects of stratification due to differences in density, salinity, temperature and oxygen content of the brine stream entering it from the SWRO plant.

After extensive studies, the authorities eventually authorised the discharge of the plant brine into Cockburn Sound, but set very tight and strict limits on how the brine had to be discharged and how quickly it had to be mixed and diluted to the same conditions as the bulk seawater around it. The SWRO plant was started in October 2006 and, to date, it appears as if the measures taken to ensure brine diffusion and mixing are meeting the criteria set for the discharge and there have been no reported adverse effects on conditions in Cockburn Sound.

In other locations the potential effects of brine disposal have been minimised or mitigated by co-locating new SWRO plants near existing power facilities and blending the brine flow from the SWRO into the existing cooling water discharge from the power plant. This way the brine is rapidly diluted while helping to reduce the temperature of the cooling water discharge slightly. It seems that in most cases, brines from SWRO plants can be discharged back into the sea with minimal apparent impact, provided adequate care is taken.

Other Opportunities for Membrane Technology

In addition to desalination, membrane technology offers the potential to improve the quality of water from other sources. Membranes can be broadly categorised by their ability to reject specific ions, molecular weight compounds or particle sizes. Generally, three families of membrane are considered: reverse osmosis (RO), nanofiltration (NF) and ultrafiltration/microfiltration (UF/MF) membranes.

RO membranes are the tightest membranes and are typically characterised by their degree of rejection of sodium chloride. Most RO membranes will reject more than 99.5% of sodium chloride applied. As stated earlier, SWRO membranes can reject 99.8%. In addition to rejecting monovalent ions, RO membranes provide a barrier to most naturally occurring and synthetic organic compounds, as well as pathogens and viruses. As an extreme example, RO membranes are utilised by the military worldwide as a key part of equipment intended to purify available water sources after any use of nuclear, biological or chemical weapons on the battlefield.

In civilian applications, RO is used to upgrade saline aquifer water. In Bahrain, for example, RO is used to treat water from the aquifers A, B, C under Bahrain. Each aquifer has its own salinity and other impurities. In other regions of the world, RO is used to remove general salinity, specific ions or compounds from otherwise unusable water sources.

Nanofiltration membranes can generally be characterised by the ability to reject a divalent salt such as magnesium sulphate. NF membranes can be used to reject divalent salts, naturally occurring organic compounds and colour components in water.

A large NF plant located at Mery-sur-Oise outside Paris is used to upgrade the water abstracted from the river Oise by reducing the concentration of total organic carbon to help to avoid the formation of trihalomethane (THM) compounds when the water is disinfected with chlorine before being put into supply. In addition to this, the NF membrane provides a barrier to any pesticides that leach into the Oise from agricultural land and also improves the aesthetic attributes, such as the hardness and colour, of the final water going into supply. NF also gives a high removal of micro-organisms, pathogens and viruses while operating at lower pressures and, therefore, lower power demand than RO membranes.

Ultra and microfiltration membranes are generally used in water treatment as a means to ensure removal of micro-organisms, such as cryptosporidium, as well as removal of high levels of pathogens and some high molecular weight compounds. UF/MF membranes are also used as a pre-treatment stage to NF or RO systems on brackish and seawater systems.

The use of membranes for desalination or for the upgrading of other water sources is not restricted to industrialised nations. Many developing nations have been among the pioneers and early adopters of the use of membranes for both seawater and brackish water treatment.

RO is a basically simple process requiring conventional pumping, piping and instrumentation equipment. The skills to install and maintain this kind of equipment can be found in most countries. RO installations also scale up fairly simply. A large or even very large RO system contains essentially the same unit processes and parts as a small unit, but on a bigger scale.

In the early days of RO development, when it was a relatively expensive process, RO was typically used in the 'developed' countries as a means of producing high-grade, high-value industrial water. However, some of the first uses of RO for drinking water were pioneered by countries like Abu Dhabi to provide reliable water supplies for remote villages and settlements around their borders with Saudi Arabia. These plants were, in fact, mounted on trailers and the pumps driven by diesel engines as heavy duty electrical power was not always available in these remote locations. As the cost of RO was reduced, then commercial applications started to develop in areas such as Sharm el Sheikh in Egypt. The development of this region for tourism led hotels to install SWRO systems to provide reliable supplies of safe, good-quality water for their kitchens and guest amenities.

One of the first, large-scale applications for RO in the treatment of municipal wastewater was developed at the Madras Refinery Limited (MRL) site in India. To ensure a reliable source of water throughout the year, MRL invested in a pretreatment plant and an RO system to take the effluent from a local municipal wastewater treatment plant. This effluent is then converted into high-grade process water that would allow the plant to operate through the year without impacting the water supplies available to the local community during times of drought or water stress.

As the capital and operating costs of membrane systems has reduced, this should open up more opportunities for developing countries and regions to use the process to augment existing water supplies.

While the cost of desalinating both brackish water and seawater has reduced significantly for any community located by the sea, suitable river or aquifer, the high cost of distributing that water to more remote locations has remained a significant and difficult problem. In many regions, the practicality or cost make laying a pipeline to take water from a large centralised water plant on the coast to a remote community inland impractical. The only way to distribute water in these areas is often by road tanker or by using bottled water for drinking. Membrane processes may, however, offer small communities, with access to low-grade-quality water the possibility to upgrade that water to something that is fit for human use, at a price that is affordable and sustainable compared to other options.

One concept is that of a 'Small Communities Packaged Plant'. The thought is that a simple, robust method of water treatment based on membrane filtration could be developed. Depending on the water source and quality available, a simple UF/MF process may be sufficient to upgrade a biologically contaminated source, such as a river or surface water, into a safe source for domestic use.

For locations where the available water sources may be high in general salinity or a specific ion, as well as being contaminated with harmful organisms, a second membrane process such as NF or RO could be used. It is thought that even today approximately 60,000 communities in China alone do not have access to a reliable supply of clean water. If regions such as the Indian sub-continent and Southeast Asia are added then there is a huge need and potential to encourage the development of a process and equipment to address that need.

Looking forwards, it seems that membrane processes will become more widespread and better accepted as a way to produce safe, economical water supplies in both the industrialised nations and the developing nations around the world. Even in 2007, the strong growth in membrane applications is generating large investments in membrane research and manufacturing capacity for the anticipated future demand.

If the past 20 years is any guide, then we should expect that the relative cost of water production will continue to fall. As that cost falls, it most likely will bring the technology into a range that can be afforded by the less industrialised and less developed nations to provide access to safe drinking water for more people. The cost reductions are likely to be more incremental changes and not the large step changes we have seen in the last 20 years as membrane performance was improved and ERD was developed.

Index

A
Abu-Zeid, K., 58
Ackerman, F., 101, 132
Acreman, M., 78
Active, Beautiful and Clean Waters (ABC) Programme, 245–246
Adaptation, anticipatory, 90
Adaptation options/measures, 188, 193
Adaptation policies, climate change, 166
Adaptation strategies, 165
Adapting Spatial Planning to Climate Change (ARK), 166
Adaptive management, 91, 92, 103, 107
Adger, N., 90, 95, 104
Administrative Procedure Act (APA), 94–95
Africa, HIV/AIDS, 203
  population, 6
  water supplies, 203
Agarwal, A., 76
Agenda 21, Rio de Janeiro, 127
Age structure, 37, 58
Aggleton, P., 206
Aging, 36, 37
Agriculture, 16, 45, 56, 72, 76, 90, 115, 150, 175, 216, 225
  Aragon, 217
  irrigation withdrawal, 70
  nitrate emission, 149
  nonpoint pollution, 154
  rainfed, 66, 68–69, 72
  water use, 1, 19, 45, 150
AGUA project (Spain), 159–160
Air pollution, 176, 179
Albiac, J., 149, 150, 159
Alcamo, J., 71
Alley, R., 90, 95
Al Sairafi, A.B., 90

Amazon, oxygen, 116–117
Anatolian peninsula, 151
Anderson, T.T., 98
Anticipatory adaptation, 90
Aquatic ecosystems, 77
Aquifers, sustainable management, 149–150
Aragon (Spain), sustainability policies, 133–134
Aragon Water Pact, 213
Aranda-Martín, J.F., 213
Arthur, R., 99
Arun II Dam, Nepal, 21
Arvai, J.L., 90, 103, 104, 106
Ashkelon desalination plant, 252
Ashton, P.J., 98, 195–197, 203, 208
Asia, population, 6
  urbanisation, 10–13
Association for Evolutionary Economics (AfEE), 130
Aswan Dam (Egypt), 4
Ataturk Dam (Turkey), 4
Australia, climate change, adaptation policies, 167

B
Baltic Sea, dying water bodies, 66
Bangladesh, deforestation/floods, 116
  population, 7
Bank of Sweden's Prize in Economic Sciences, 135
Barnett, T., 195, 197, 198, 202–204, 207, 208
Baylies, C., 207
Bebbington, J., 132
Berndes, G., 69, 70
Bhakra Nangal, India, 4
Bharat, S., 206
Biesbosch, freshwater tidal zone, 189–190

Biocapacity, overshoot, 77
Biodiesel, 19
Biodiversity, 78
Bioenergy, water requirements, 67, 69
Biofuel, 19, 78, 216
   crops, water demand, 20
Biomass production, green water, 67
Biotechnology, 3–4, 25
   advances, 3–4
Biswas, A.K., 1, 4, 11, 32, 89, 94, 106
Blann, K., 93
Block, W.M., 99
Bloom, D.E., 57
Bloxham, M.J., 105
Blue water, 67, 81, 245
BOD, European rivers, 153
Bolton, R., 207
Bormann, B.T., 99, 100, 104, 105
Bosworth, B., 155
Brackish groundwater seepage, 175
Brazil, economic growth, 4
   forests, 116–117
BRICs, economic growth, 4
Browder, G., 55
Bruch, C., 89, 95–96, 98, 102
Brundtland Report, 127
Brunner, R.D., 99
Bullock, A., 78
Burton, I., 90, 104

C

Cadmium, European rivers, 153
Calder, I.R., 80, 81
Campbell, C., 205
Cannon, J.Z., 94, 98
Capacity building, 142, 144–146
   diary, 139
Carden, K., 99, 102, 104
Catchment, importance, 79, 80
   water movement, 81
Cauvery river, water shortage, 73
Change, challenges, and chances ('cha-cha-cha'), 139
Chaos Theory (Gleick), 91
Cheek, K.A., 102
Chen, B.S., 52
Chetty, D., 199–200
China, aged population, 40
   cities, 54
   deforestation/floods, 116
   drinking water/wastewater management, 13
   economic growth, 4–5
   GDP, 5
   industrial wastewater reduction, 76
   population, 7
   urban centres, wastewater treatment, 55
   water, aging and urbanisation, 57
Chromium, European rivers, 153
Chye, K.T., 237
Cities, classification, 54
City wastewater, 75
Clark Fork Project, 97–98
Clarke, A., 104
Climate change, 26, 49, 58, 66, 86, 89, 90, 105–107, 115–116, 133, 149, 165
   adaptive water management, 100, 106
   European Union Climate Change Committee (CCC), 17
   Intergovernmental Panel on Climate Change, 103
   Kyoto Protocol (climate change), 126
   Mediterranean, 151
   The Netherlands national strategy, 99
Climate-proofing, 165, 172
Climate scenarios, 167–168
Cloud forests, harvesting fog water, 78, 79
Club of Rome, 51, 74
CO_2 reduction, 133
Coastal defences, 172
Colborn, T., 74
Coleman, W.T., 94
Columbia River, adaptive management, 98
Colvin, C., 205
Combes, S., 90
Communities of practices, 147
Comprehensive management, 121
Congo, population, 7
Connolly, W.E., 126
Consumer preferences, 129
Continuing education, 145
Cooney, R., 99, 104
Copper, European rivers, 153
Cornell, M., 195
Cornish, G., 155
Corporate social responsibility, 126
Corruption, 13, 27, 29, 56, 90, 121
Cost-benefit analysis, 131
Cost recovery, 155
'Crop-per-drop', 67
Crop yields, 69

D

Dalton, M.G., 60
Dams, 21, 95, 132, 230
   Arun II (Nepal), 21
   Aswan Dam (Egypt), 4
   Ataturk (Turkey), 4
   Glen Canyon, 98

irrigation, 155
licenses, 97
Marina Barrage, 239–240
Nagara Barrage (Japan), 21
Pelton Round Butte Project, 98
St. Lawrence-FDR Project, 98
Sardar Sarovar (India), 21
sustainability assessment, holistic methods, 134
Tehri (India), 21
Da Vinci, Leonardo, 1
Deforestation, 82, 116
Deming model of quality management, 139
Democracy, comprehensive management of water resources, 121–123
Demography, 53
Denmark, climate change, adaptation policies, 167
de Pater, F., 188
Depoldering, 190
Desalination, 4, 25, 27, 119, 156, 251–258
  silt production, 119
Desalination plant, Ashkelon, 252
Desert Rock Energy Project, 19
Detergents, 152
Deutsches Institut für Wirtschaftsforschung, 135
Development dialogue, alternative perspectives, 128
Dhaka, drinking water/wastewater management, 11
Dhar Chakraborti, R., 58
Diamond, J., 65, 71
Diné Power Authority, 19
Disease agents, reduction, 75, 85
Diseases, 28, 179, 180, 195, 206
  food-related, 179
  HIV/AIDS, 4, 57, 61, 195–210
  infectious, 180
  Lyme, 180
  malaria, 179, 198, 207
  public health, 179–180
  tuberculosis, 198
  waterborne, 28–29, 45, 196, 204
Distribution networks, 157
Doremus, H., 91, 94, 99
Dorrington, R.E., 200
Downward, S., 161, 162
Dragusanu, R., 8
Drainage basins, biodiverse, 81
Drakakis-Smith, D., 50
Drivers of change, 3
Droughts, 17, 24, 44, 159, 166, 170, 215
  deforestation of upstream areas, 116

Duda, A.M., 79
Dutch coastal defences, 172

E
Eberstadt, N., 201
Ebro, 153, 222, 223, 233
  basin, 216, 222
  delta, 215
  water transfer, 158
Economic development, 43
Economic growth, 4–6, 128
  rates, 5
Economic history, 134
Economic Man, 127, 131
Ecosystem management, 85
Ecosystem protection, 76–77
Ecosystem services, 77–78
Education, 28, 43, 50, 121, 133, 186
Electricity generation, 6
Electricity requirements, 2
El Feki, S., 201
Energy, 15–17, 45, 175–176
Energy recovery devices (ERD), 252
Environmental charges, 126
Environmental concerns, 21
Environmental flow, 72
Environmental Impact Assessment (EIA), 126
Environmental management systems, 126, 132–133
Environmental taxes (green taxes), 126
Ergüden, S., 35
Erskine, S., 195, 197, 203
Escaut, nitrates/phosphates, 152
Ethanol, biofuel, 19
Ethiopia, population, 7
Etzioni, A., 127
European Environment Agency (EEA), climate change, adaptation policies, 166
European rivers, water quality, 153
European Union Climate Change Committee (CCC), 17
European Union, neoclassical economics, 133
Europe, population, 7
  aged, 38
  southern, water scarcity, 151
  water, aging and urbanisation, 58
Evaporation, 67, 78
Everglades, adaptive management, 98
Evolutionary economics, 130

F
F4 (factor four) concept, 51
Fair trade, 129
Falkenmark, M., 65–86

Farber, D.A., 98, 100
Feldman, I.R., 107
Fertilisers, 76, 231
Fertility rates, 7
Finland, climate change, adaptation policies, 167
Floods, 24, 44, 81–82, 115, 155, 216
  deforestation of upstream areas, 116
  protection, 177
Florida Everglades, adaptive management, 98
Flournoy, A.C., 100
Food, 68–70
  demands, 30
  import, 57
  losses, 14
  markets, 53–54
  relation to water requirements, 1, 14
  supply, 54, 214
  water requirement, 68–69
Food and Agriculture Organisation (FAO), 216, 234
Food availability, 14
Food chains, 66, 174
  response to climate change, 174
Food consumption, 68
Food-exporting countries, 54
Food production, 44, 77, 115, 167, 216
  increase, 19, 44, 45, 56, 216–217, 227
  water requirements, 68–70
Food requirements, global, 1, 68–70
Food safety standards, 180
Food security, 4, 22, 23, 30, 44, 58, 138, 217, 225
Food shortage, 234
Food web, pollution, 74
Ford, D., 131
Forest coverage, decrease, 115
Forest resources, 94
  management, 116–117
Forestry, 80, 166
  management, 99, 186
Forests, acid rain, 21
  deforestation, 82, 116
  groundwater recharge, 78
  oxygen, 116–117
  protection, 79, 133–134
  reforestation, 82
  upland, 78
  water resources, 117
  water retention, 186
Fredericksson, J., 203
Freeman, J., 98
Free trade, 3, 133
Friedman, S.R., 196
Fullbrook, E., 134
Fusfeld, D.R., 134

G

Garaway, C., 99
Garduño, H., 84
Genthe, B., 205
Geographic Information Systems (GIS), 138–139
George, S., 129
Gilbert, L., 202, 207
Giroux, H., 129
Gleick, J., 89, 91
Gleick, P.H., 89, 100, 207
Glen Canyon Dam, 98
Global environmental movement, 21
Globalisation, 3–4, 54
Governance, 28
  concepts, 83–84
Grabs, W., 27
Great Lakes, adaptive management, 98
Greene, D.M., 99
Gross domestic product (GDP), 5, 16, 27, 52, 132, 178, 246
Green labelling, 126
Green water resource, 66–70, 83, 85
Gross, J.M., 96
Groundwater, brackwater seepage, 175
  control, 160
  levels, 174
  mining, 127
  mismanagement, Spain, 159
  nitrate-rich, South Africa, 205
  overuse, 14–15
  pollution, 75, 216
  pumping, 14
  recharge, 78, 231
    protection, 78
  salination, 173
  seepage, 79
  surface runoff/percolation, 69
Groundwater tables, declining, 15
Guadalquivir, heavy metals, 152
  nitrates/phosphates, 152
Guadiana basin, 159
Gulf of Mexico, dying water bodies, 66
Gunderson, L., 94

H

Hai river, water shortage, 73
Hamman, R., 98
Hanmer, R., 103
Hansen, L.J., 90, 91
Harman, W., 132

Harremoës, P., 127
Heavy metals, 76, 149, 152, 161
Heinzerling, L., 132
Heiskanen, A.S., 81
Hernández, N., 158
Hewett, P., 54HIV/AIDS, 4, 61, 195
  clean water/sanitation, 203
Hoagland, R.A., 103
Holland, R., 81
Holling, C.S., 91, 93
Hooper, E., 197
Hornstein, D.T., 91, 101
Houck, O.A., 96
Household size, water consumption, 52
Housing protection, 177
Huai river, water shortage, 73
Human demographics, 35
Human Development Index (HDI), 46–48
Human resource development, capacity building, 137–148
Hunger alleviation, water requirements, 68
Hunter, D.J., 107
Hydrocide, 74
Hydropower dams, licensing, 97–98
Hydropower generation, 2, 17, 67

I
Iberian peninsula, 151
Ideological orientation, 129
IJssel, 191
IJsselmeer, 173–174
Immigration, 3–4
Income, middle-class, 7–8
India, economic growth, 4–6
  population, 7
  tanneries, 76
  water, aging and urbanisation, 58
Indigenous people, forests, 117–118
Indonesia, discharge of water wastes, 120
  Environment Center, 120
  violent conflicts, 117
Industrial pollution, 45, 76
Industrial revolution, 50
Industrial water use, 2, 14, 151
Infectious diseases, 180
Information and communication services, 10
Information technology, 53
Institutional economics, 128–129
Institutional expertise, 9
Integrated water resources management (IWRM), 94, 146
International Confederation of Associations for Pluralism in Economics (ICAPE), 134, 135

Internet, direct democracy, 121
InWEnt, 144–146
Irrigation, 66–67, 69, 74, 84, 119
  water use, 149, 155, 213–234
Irrigation water, price, 155
Italian peninsula, 151

J
Jakarta, drinking water/wastewater management, 11
Janse van Rensburg, E., 197
Japan, institutional expertise, 9
  median age, 8
  Nagara Barrage, 21
  urban water infrastructure, 12–13
Jayne, T.S., 204–205
Jeans, E.G., 27
Jeffrey, P., 94
Johannesburg Plan of Implementation, 45
Júcar river, 159, 218

K
Kabat, P., 90, 94, 102
Kahan, J.H., 107
Kamminga, E., 195, 196, 203, 204, 206–209
Kampen/IJsseldelta study, 190–191
Kanabus, A., 199, 203
Karachi, drinking water/wastewater management, 11
Karkkainen, B.C., 93, 94, 98, 101–103
Keilman, N., 52
Kelley, L.M., 201
Kennedy, John F., 1
Kgalushi, R., 204, 209
Khoo, T. C., 237
Kippen, R., 60
Knowledge/innovation, access, 148
Korten, D., 130, 132
Kranz, N., 90, 94, 96, 102, 103Kras, E., 128
Kuppusamy, B., 201, 202
Kuznetz curve, 120
Kwadijk, J., 171
Kyoto Protocol (climate change), 126

L
Labour productivity, 51–52
Lakes, 78–79
  Great Lakes, 98
  IJsselmeer, 173
  Lake Michigan, 19
Landscape components, 78
Land use, 82
Lang, A.T.F., 99, 104
Lannerstad, M., 71, 73

Lasage, R., 165
Latrines, 75
Lawrence, J., 91
Lawrence, P., 45–48
Laws, adaptive management, 94–96
　periodic revision, 94
　water management, 93, 94
Lawson, R.L., 16
Lead, European rivers, 153
Leadership, 122, 137
Leadership problems, 139
Lee, K.N., 91, 98
Legal frameworks, change, 101
Lenton, R., 27
Leusink, A., 99, 165
Lewin, R., 89
Life-cycle analysis (LCA), 126
Life expectancy, 7
Light, S.S., 93
Limpopo, water shortage, 73
Liquid water/blue water, 68
Liu, J., 52
Logging, national/local interests, opposition, 117
Lomax, I., 251
London, growth, water and wastewater infrastructure, 11
Lundberg, M., 207
Lundqvist, J., 54, 74
Lyme disease, 180

M
McDonald, P.M., 60
MacKay, H.M., 89, 103
Mälardalen (Sweden), sustainability policies, 133
Malaysia, 240
Mangrove forests, 78
Man–land–water waste system, 80–81
Marais, H., 195–197, 203
Marañón-Pimentel, B., 28–29
Marina Barrage, 239–240
Market fundamentalism, 129
Markets, 129
　supply and demand, 129
Martínez, Y., 159
Meadows, 78
Means, E.G., 60
Medema, W., 94
MENA water programme, 36, 58, 146
Mental maps, 128
Meretsky, V.J., 99
Methodical competence, 147
Michel, B., 199–200

Middle East, water, aging and urbanisation, 58–59
Migration/emigration, 7
Mijares river, 233
Milly, P.C.D., 92
Mismanagement, 160
Missouri River, 18–19, 95
Mitchell, C., 99, 103
Moir, W.H., 99
'Monetary reductionism', 132
Monitoring, 45, 91, 93, 141, 184, 187, 249
Montgomery, M., 54
Montreal Protocol (ozone layer), 126
Montville, J., 105
Mortality, heat stress, 179
Mujeeb, S., 206–207
Murray-Darling river basin, 97
Murua, R., 149
Myer, L., 197

N
N11 countries, middle-income class, 7
Nagara Barrage, Japan, 21
Nanofiltration (NF), 257
Narain, S., 84
Nardi, P.M., 207
National Energy Technology Laboratory (NETL), 18
National Environmental Policy Act (NEPA), 94
National parks, 79
National Programme on Adapting Spatial Planning to Climate Change (ARK), 166
National Water Agency, 141
Natural disasters, 24
Needs assessment, 147
Neoclassical economics, 126, 130
Neo-liberalism, 129
Nepal, deforestation/floods, 116
Nepotism, 13
Netherlands Environmental Assessment Agency (MNP), 173
Netherlands, reaction to climate change, 99, 165–193
　Royal Netherlands Meteorological Institute (KNMI), 167
　temperature increase, 168
Neuman, J.C., 91, 95, 101, 103, 105
New York, growth, water and wastewater infrastructure, 11
Nigeria, population, 7
Nitrates, 152, 153
　HIV/AIDS, 205
Nitrogen surplus, soils, 154

Nonlinear dynamics, 91, 95
Non-monetary impacts, 127
Nonpoint pollution control, 149–150
North Africa, water, aging and urbanisation, 58
Northwest Forest Plan (U.S.), 100
Norton, B.G., 102, 103
Nuclear power, 2, 15
Nutrient pollution, 76, 149

O

Obi, C.L., 204, 207
OECD, pollution levels/decrease, 5
Oise river, 257
O'Neill, B.C., 52
Oregon Plan for Salmon and Watersheds (Oregon Plan), 99
Organisational development, 145
Oxygen supply, forests, 116–117

P

Pahl-Wostl, C., 89, 94, 101–104
Pakistan, population, 7
Parma, A.M., 99
"Passive" adaptive management, 93
Peacebuilding, importance of forests, 118, 120
Peltzer, K., 199–200
Percival, R.V., 96
Permits, 93
Perry, C., 155
Personnel development, 142, 144
Peter Principle, overtaxation, 138
Pfeifer, H., 137–148
Phaswana-Mafuya, N., 199, 206, 208
Phosphates, 149, 152
Physical water scarcity, 70
Pittock, B., 97
Pittock, J., 81
'Plan-do-check-act'-circle, 139
Platte River, adaptive management, 98
Polder conversion, 190
Policies, 93
Policies, implementation, 1, 5, 15, 19, 24, 27, 28, 45, 82, 90, 93–98, 100–106, 117, 121, 126, 149, 156, 160, 181, 192–193, 214, 230
Political Economic Person (PEP), 129, 131
Political system, 148
Pollution abatement imperative, 85
Pollution control programmes, 84
Pollution, decrease, 5
   persistent, 75
Pollution fees, point sources, 83–84
Poole, G.C., 93
Population, 3
   age distribution, 8, 37
   child births, 36
   deaths, 37
   expansion, 37
   global total, 7
   growth rates, decline, 6
   increase, Asian megacities, 12
   predictions, uncertainties, 60–61
   urban, 10
Porsuk, heavy metals, 152
Positional analysis (PA), 131, 132
Positivism, science, 128
Post-Autistic Economics network, 134
Post-conflict consolidation, 122–123
Potential Support Ratio (PSR), 38
Potter, A., 205
Poverty, alleviation, 4
Power generation, water use, 151
Precipitation, 67
Precision watering, Israel, 119
Privatisation, 138
Professional knowledge/training, 146, 147
Profeta, T.H., 95
Profit maximisation, 129
Public relations/public awareness, 147

Q

Quirk, P.J., 102

R

Rahaman, M.M., 94
Rainfed agriculture, 66
Rainwater harvesting, 66
Rainwater runoff, prevention, 70
Ralph, S.C., 93
Ramasar, V., 195–197, 203, 204, 208
Ramsar Convention, 79
Rangachari, R., 4
Rascher, J., 195–210
Recreation sector, climate change, 180
Recycling, water, 58, 119, 247
Reforestation, 82, 117, 118
Regional cooperation, 147
Regional Water Agency, 141–142
Regulations, 93
Reid, G., 195
Religion, indigenous, relation with tropical forests, 118
Resilience, 172
Resource adaptation, 85
Resource productivity, 52
Resource-use impacts, household size, 52
Resource utilisation, 5
Reutilisation, 157

Reverse osmosis (RO), 251–258
Rijsberman, F.R., 70
Rio, Agenda, 127
Rio Declaration, oxygen issue, 117
River basins, Europe, 161
    management, 122
    management units, 84
River depletion, 66
River discharges, 99
River flow, 83
    control, 115
River levels, 176
River parliaments, India, 84
River patrols, 245
River pollution, 66
    farming, 231
    wastewater treatment, 66
Rivers, agricultural leaching, 23
    Cauvery, 73
    Columbia, 98
    Dommel, 191
    Ebro, 222, 223
    European, water quality, 153
    Guadiana, 159
    Hai, 73
    Huai, 73
    IJssel, 191
    Mijares, 233
    Mississippi, 23
    Missouri, 95
    Murray-Darling, 97
    nitrate, 149
    Oise, 257
    peak discharges, 173
    Rhine, 173, 241
    Segura (Spain), 159
    Singapore, 237
    Sur (Spain), 159
    Tajo, 152, 218
    transnational, 66
    Turia, 233
    Yellow, 73, 82
Rockström, J., 66, 68–70, 83
Rothenberg, L.S., 106
Routeplanner, The Netherlands, 165
Ruhl, J.B., 89, 91, 92, 94, 95, 98, 99, 101
Runoff, prevention, 70
Rural areas, family systems, 9–10
Rural population, 40
Russia, economic growth, 4

S
San Francisco Bay-Delta, adaptive management, 98

Sardar Sarovar dam, India, 21
Savings, population growths and productivity gains, 5
Sawtooth National Forest, 95
Scholz, J.T., 102, 103
Schueller, S.K., 93
Schuringa-Wegelin, M., 195, 196, 203–210
Science and technology, sustainable management, 118
Sea level rise, 170
Seawater desalination, 251–258
Seawater reverse osmosis (SWRO), 251–252
Seely, H., 92
Segura basin (Spain), 159
Seine, heavy metals, 152
    nitrates/phosphates, 152
Sendzimir, J., 94
Sewage treatment, 89, 152, 174, 209
    secondary, 152
    tertiary, 152
Shaver, C., 99
Shindler, B., 102
Simonovic, S.P., 74, 75, 85
Singapore, 237
Smakhtin, V., 66, 72
Smith, C.L., 99
Snrech, S., 56
Social irrigation, 214
Söderbaum, P., 125, 130, 132, 135
Söderlund, L., 57
Soil moisture/green water, 68
Soils, nitrogen surplus, 154
Spain, irrigation, 213
    water management, 159
Stakeholder involvement, 102
Stern Review Report, 177, 188
Stiftel, B., 102
Strategic Human Resource Development (SHRD), 144
Streamflow, 78–79
    residual, 72, 79
Sufian, S., 197, 201, 202
Suloway, J.J., 98
Sunter, C., 195, 197
Sustainability assessment, 130
    holistic methods, 134
Sustainability economics, 135
Sustainable development, 125
Sustainable management, 149–150
System level development, 144

T
Tajo, 218
    heavy metals, 152

Takahashi, K., 115–124
Tanneries, 76
Tarlock, A.D., 95, 99, 100
Tax accumulation, 51, 53
Taylor, R., 161
Tehri dam, India, 21
Terrestrial ecosystems, 77
Thames, nitrates/phosphates, 152
Thermal power, 2, 15
Thermoelectric generation, 18
Third World Centre for Water Management, 4
Thomas, G., 85
Thomas, J.W., 99, 100
Thrower, J., 95, 100, 102
Tilburg, 191–192
Tladi, B., 206
Todaro, M., 51
Toilets, 75
Tortajada, C., 1–32, 54
Training capacities, 147
Transboundary basins, 82
Transport sector, 176
Trihalomethane (THM), 257
Turton, A., 195

U
Ultrafiltration/microfiltration (UF/MF), 256
United Kingdom, climate change, adaptation policies, 166
United Nations Conference on the Human Environment, Stockholm (1972), 21
United States, population, 7
U.N. Peacebuilding Commission, 122
Upland forests, 78
Urban centres, classification, 55
Urban population, growth, 41
Urban Wastewater Treatment Directive, 152
Urban water infrastructure, Japan, 12–13
Urban water use, 150
Urbanisation, 3, 10, 35, 40, 51–58, 138
U.S. Clean Water Act, 96

V
Vakkilainen, P., 57
van Drunen, M.A., 165–194
Van Dyk, A., 195
van Ginkel, H., 102
van Ierland, E.C., 185, 187
van Rensburg, E.J., 197
van Schaik, H., 90, 102
Van Wijk, C., 196, 203–208
Varis, O., 9, 35, 57–62, 89, 94
Veraart, J., 172
Volkman, J., 98

von Weizsäcker, E.U., 51

W
Wailand, W.J., 101, 106
Walker, L., 195
Walters, C., 92, 93, 103
Warner, J., 83
Wastewater, 2, 9, 23, 68, 75, 84, 152, 237
  "black-holes", 11
  city, 75
  disposal, 2
  households, 22
  increase, 152
  industrial discharges, 22
  industrial, lack of secondary/tertiary treatment, 23
  nutrient reclamation, 75
  reuse, 27
  safe disposal, 12
Wastewater dischargers, pollution control programmes, 84
Wastewater disposal, water pollution, 74
Wastewater management, developing countries, 24
  Singapore, 237
Wastewater problem, megacities, 11
Wastewater reclamation/recirculation, 75, 76
Wastewater requirements, elderly, 10
Wastewater treatment, 23, 25, 49, 55, 74, 152, 248
  advances, 25
WATECO, 157
Water authorities, 157
Water availability, carrying capacity, 71
  smallholder farming, 70
Water contamination, 22
Water crowding, 71
Water distribution networks, 157
Water efficiency, 67
Water Framework Directive, 154
Water governance, 83
  adaptive, 103
Water institutions, 157
Water laws, 90
  adaptive management, 99, 106
  Spain, 160, 214
  U.S. Clean Water Act, 96
Water management, elderly population, 9
  proactive, 80
Water planning and management, 2
Water pollution, 22, 45, 65, 74, 76, 84, 229
  abatement, 85
  agricultural, 76, 154
  agricultural chemicals, 20

avoidance, 74
Code of Good Farming Practice, 229
control programmes, 84
developing countries, 22
dilution, 74
domestic, 75
fees, 83–84
fertilisers, 154
food web, 74
groundwater, 75, 216
industrial, 45, 76
nitrates, 154
nonpoint sources, 22, 23
point sources, 96
prevention, 21
reduction, 152
scale, 74
Singapore, 239
soil erosion, 20
subsurface waters, 154–155
Water pollution fees, point sources, 83–84
Water Poverty Index (WPI), 45–48
Water pricing, 155, 156, 162
Water problems, abstract theories, 134
Water productivity, 69
Water projects, funding, 21
  planned, 3
Water pumping, 2, 15–16
Water quality, 1, 67, 120, 149
  degradation, 67
Water reform processes, leadership, 137–148
Water resource adaptation, 85
Water resource management, 115
Water scarcity, economical, 71
  good management, 70
  physical, 70
Water sector, energy use, 20
  investment, 6
Water shortage, 71–75, 85, 190, 233
  chronic, 71
Water, source of divinity, 123

Water stress, 71, 149
Water technologies, adaptation, 58
Water transfer projects, sustainability assessment, holistic methods, 134
Water use, agricultural, 1, 16, 45, 150
  industrial, 2, 45, 150
  irrigation, 150, 155, 213
  power generation, 151
  reduction, 150
  by sector, 150–151
  urban, 150
  wastefulness, 67
Water/wastewater infrastructure, 11
  natural disasters, 24
Water/wastewater management, megacities, 11
  Singapore, 237
West Africa, agriculture, 56–57
  water, aging and urbanisation, 56
Wetlands, 78
  special value, 79
Whiteside, A., 195, 197, 198, 202–204, 207, 208
Wiener, J.B., 100
Williams, A., 201
Williams, B.K., 99
Williamson, J., 57
Wilson, D., 8
World Commission on Dams, multiple criteria approach, 132
World economy, 53–54
Wratt, D., 97

Y

Yellow river, water shortage, 73, 82

Z

Zaelke, D., 90
Zeeland, 173
Zetoun, M., 83
Zuid-Holland, rivers, 173